Electric Motor Maintenance and Troubleshooting

Second Edition

Augie Hand

New York   Chicago   San Francisco
Lisbon   London   Madrid   Mexico City
Milan   New Delhi   San Juan
Seoul   Singapore   Sydney   Toronto

The *McGraw·Hill* Companies

Cataloging-in-Publication Data is on file with the Library of Congress

McGraw-Hill books are available at special quantity discounts to use as premiums and sales promotions, or for use in corporate training programs. To contact a representative, please e-mail us at bulksales@mcgraw-hill.com.

Electric Motor Maintenance and Troubleshooting, Second Edition

1 2 3 4 5 6 7 8 9 0   DOC DOC   1 0 9 8 7 6 5 4 3 2 1

ISBN   978-0-07-176395-0
MHID   0-07-176395-3

Sponsoring Editor Roger Stewart	**Copy Editor** William McManus	**Composition** TypeWriting
Editorial Supervisor Janet Walden	**Proofreader** Paul Tyler	**Art Director, Cover** Jeff Weeks
Project Editor Patricia Wallenburg	**Indexer** Karin Arigoni	**Cover Designer** Mary McKeon
Acquisitions Coordinator Joya Anthony	**Production Supervisor** George Anderson	

To my wife Leanne, whose skills in sentence structure have contributed greatly to my books.

And to my good friend, Ed Frisbee. His expertise in the electric motor field has helped immensely in updating this book.

About the Author

Augie Hand has over 40 years of experience in the electric motor field. He owned and operated an electric motor repair shop for 15 years and originated and taught the Electric Motor Repair course for 8 years at Southwestern Technical Institute in Jackson, MN. Mr. Hand has been presenting seminars to electric motor repair shops and large industrial firms for over 15 years. He is also the co-author of *Electric Motor Repair*, 3rd ed., which is considered the industry "Bible."

Contents

Troubleshooting Single-Phase Motors

Troubleshooting Three-Phase Motors

Troubleshooting
the DC Machine

Introduction

Worldwide competition has forced big changes in industry. Predictive and preventive maintenance programs have replaced reactive maintenance. These programs are important factors in the profit and, in numerous cases, the survival of many industries. The cost per hour of downtime clearly illustrates the need for such programs.

This book explains electric motor theory and troubleshooting techniques. Its intent is to inform the technician in direct language without needless math and cross-references.

Effective electric motor maintenance and troubleshooting require a complete understanding of a motor's internal structure. The electric motor theory in this text is directly relatable to maintenance and troubleshooting. The math and motor theory are not meant to be used as exact formulas for designing or redesigning a motor or an electrical system. Instead, they are directly applied to motor problems and solutions.

The book covers all types of AC and DC motors. DC motor and generator operation (and components) will be explained first, followed by AC single-phase motors, and then three-phase motors. Connections and their numbering systems are included with the description of each machine.

Two troubleshooting procedures will be presented. The first procedure tests the machine's components to quickly see if it should be removed for repair. The next procedure locates the problem inside the machine.

Although electric machine problems can be difficult, there is no problem that does not have an explanation and a solution. It's just a matter of gathering all the facts, and applying the appropriate logic.

Additional Resource for Instructors

The answers to the test questions at the end of each chapter are available to educators online from McGraw-Hill Professional. Instructors should contact their McGraw-Hill sales representative for access information.

If you need to locate your McGraw-Hill sales representative, go to www.MHHE.com and click "Find My Sales Rep."

Chapter 1

The DC Machine

The DC machines described in this chapter are motors or generators. The term *machine* is used when the test procedure is the same for both. A motor can be converted to a generator (and vice versa), so the terms *motor* and *generator* are used only when the explanation applies to one or the other.

DC generators produce very high-quality power, but (because of maintenance and other costs) AC powered DC drives are used for most motor applications.

DC power is used for some high-voltage power transmission. DC current flows through the whole area of a wire, making it more efficient than AC for long distance transmission. High-voltage AC current uses only the outer portion of the wire. Only two transmission lines are needed with DC, and in an emergency, the earth can be used as a second conductor. At the user end, DC is converted to three-phase power and distributed to substations.

DC generators also make the best power for arc welders. They provide steady voltage and a non-fluctuating current that flows in one direction.

The DC motor has excellent speed control with very good torque and horsepower characteristics. Because of its armature design and function, it has very smooth torque from 0 RPM to base speed. The DC motor also has full-rated horsepower above base speed.

Basic Electricity as It Applies to Motors

The properties of electricity are volts, resistance, and amperes. Voltage is the driving force, resistance is the work to be done, and amperes get the work done.

Volts

Voltage (electromotive force, or emf) is the driving force that causes the amperes to flow through the resistance of the load. Even if there is no circuit or path, voltage can be present. The volt can be compared to air or water pressure. When voltage is raised, more amperes will flow through the resistance (load). When voltage is lowered, fewer amperes will flow. When the voltage is varied, the number of amperes flowing through a given resistance (load) will go up or down *with* the voltage change.

Another comparison of the volt to air or water pressure is containment. Higher voltage requires stronger (thicker) insulation.

Resistance

Resistance controls the number of amperes that flow in a circuit. When a constant value of voltage is applied, as resistance goes up, amperes go down and as resistance goes down, amperes go up. As the resistance value varies, the number of amperes varies the *opposite* way. All loads have some form of resistance. The resistance of a device is measured in ohms. (Ohm's Law is discussed later in the chapter.)

Resistance is opposition to ampere flow and measured in ohms. It is seldom measured, so is just called resistance. It is common to measure amperes.

Two factors furnish resistance to current flow in an electric motor. First is the resistance of the wire in the coils that form the poles. Each wire size has a resistance value per 1000 feet at a given temperature. Coils of wire used in the shunt field of a DC motor have enough feet of wire to limit the amperes to a safe level (and not overheat). The second factor is the interaction of the winding conductors and the magnetic circuit of the motor. This will be explained under "Counter emf" in DC motors in the section "Counter-voltage" and in Chapter 3 under "Inductive Reactance" in AC motors.

Amperes

The *ampere* is a measurement of the number of electrons flowing in a wire. The number of amperes flowing in a circuit is controlled by two factors: the voltage applied and the resistance of the load. The voltage and/or the resistance are varied to control the amperes. The formula called Ohm's Law (described later in the chapter) calculates the number of volts and/or the amount of resistance needed to predict the number of amperes in a circuit.

Most electrical breakdowns involve ampere flow. When insulation breaks down, heat created by ampere flow destroys it. Excessive amperes flowing in a wire cause a wire to become hot.

The number of amperes flowing through a coil controls the coil's magnetic strength. As the number of amperes changes, the coil's magnetic strength will vary *with* the change.

The direction of ampere (current) flow determines the polarity of the coil. Figure 1.1 shows the left-hand rule for determining the polarity of a DC coil.

The right wire size is a very important part of motor design. The wire size is determined according to its cross-sectional circular mil area. The number of amperes that flow in a motor's circuits and the motor's cooling ability determine the wire size.

The coils used in the shunt field of a large DC motor have much larger wire size (circular mils per amp) than the coils used in single- or three-phase induction motors. This is because the coils have a large mass and do not cool easily. It's common to find 1000 (or more) circular mils per amp in the shunt field coils of a large DC motor. It is also common for single- and three-phase motors to have 300 to 350 circular mils per amp. Table 1.1 shows the wire size converted to circular mils plus other data.

Additional Information on Copper Wire

This wire table can be remembered very easily if a few simple points are kept in mind:

- A wire three sizes smaller than another wire has half the area of the larger wire. For instance, No. 20 AWG copper wire has half the area of

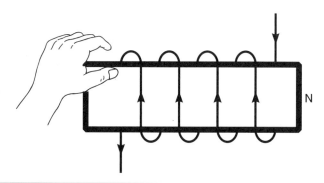

FIGURE 1.1   Left-hand rule. Place the left hand on a coil of wire with the fingers pointing in the direction of current flow. The thumb points to the north pole.

TABLE 1.1   Wire Size Conversion

AWG	Diameter, Inches	Circular Mils	Pounds per 1000 ft	Ohms at 68°F per 1000 ft
0000	0.4600	211,600.0	640.5	0.0490
000	0.4096	167,800.0	507.9	0.0618
00	0.3648	133,100.0	402.8	0.0779
0	0.3249	105,500.0	319.5	0.0982
1	0.2893	83,694.0	253.3	0.124
2	0.2576	66,370.0	200.9	0.156
3	0.2294	52,630.0	159.3	0.197
4	0.2043	41,740.0	126.4	0.248
5	0.1819	33,100.0	100.2	0.313
6	0.1620	26,250.0	79.46	0.395
7	0.1443	20,820.0	63.02	0.498
8	0.1285	16,510.0	49.98	0.628
9	0.1144	13,090.0	39.63	0.792
10	0.1019	10,380.0	31.43	0.998
11	0.09074	8,230.0	24.92	1.260
12	0.08081	6,530.0	19.77	1.588
13	0.07196	5,170.0	15.68	2.003
14	0.06408	4,107.0	12.43	2.525
15	0.05707	3,257.0	9.858	3.184
16	0.05082	2,583.0	7.818	4.016
17	0.04526	2,048.0	6.200	5.064
18	0.04030	1,624.0	4.917	6.385
19	0.03589	1,288.0	3.899	8.051
20	0.03196	1,022.0	3.092	10.15
21	0.02846	810.1	2.452	12.80
22	0.02535	642.4	1.945	16.14
23	0.02257	509.5	1.542	20.36
24	0.02010	404.0	1.223	25.67
25	0.01790	320.4	0.9699	32.37
26	0.01594	245.1	0.7692	40.81
27	0.01420	201.5	0.6100	51.47
28	0.01264	159.8	0.4837	64.90
29	0.01126	126.7	0.3836	81.83
30	0.01003	100.5	0.3042	103.2
31	0.00892	79.70	0.2413	130.1
32	0.00795	63.21	0.1913	164.1
33	0.00708	50.13	0.1517	206.9
34	0.00630	39.75	0.1203	260.9
35	0.00561	31.52	0.09542	329.0
36	0.00500	25.00	0.07568	414.8
37	0.00445	19.83	0.0601	523.1
38	0.00396	15.72	0.04759	659.6
39	0.00353	12.47	0.03774	831.8
40	0.00314	9.888	0.02990	1,049.0

No. 17 AWG copper wire. Therefore, two No. 20 wires in parallel have the equivalent area of one No. 17 wire.

- A wire three sizes smaller than another wire has twice the resistance of the larger wire.
- A wire three sizes smaller than another wire has half the weight of the larger wire.
- A No. 10 AWG copper wire is approximately 0.10 inch in diameter, has an area of approximately 10,000 circular mils, and has a resistance of 1 ohm per 1000 feet.

If there are too few circular mils per amp, the coils will overheat and the motor's insulation will deteriorate prematurely. Excessive heat increases copper loss and lowers the motor's efficiency. Copper gains resistance as its temperature rises. As copper resistance goes up, the amperes go down, lowering the motor's horsepower output.

Ohm's Law

The relationship of voltage, amperes, and resistance is explained with a formula called Ohm's Law. The following formulas, in which E (electromotive force) = volts, I (intensity of current) = amperes, and R = resistance, predict the results when designing electrical devices:

Volts divided by resistance equals amperes ($E \div R = I$)
Volts divided by amperes equals resistance ($E \div I = R$)
Amperes multiplied by resistance equals volts ($I \times R = E$)

Varying the voltage or the resistance controls amperes.

Watt

The word *watt* is short for joules per second. A watt is the measurement of power being used to do work. The number of watts is found by using the formula volts $\times$ amperes = watts. A power meter (which determines power cost) multiplies volts $\times$ amps and measures the time involved. The cost is determined by the kilowatt hour (1000 watts for 1 hour).

A motor converts electrical energy directly into mechanical energy. One horsepower (hp) equals 746 watts. *One horsepower has the ability to lift 550 pounds 1 foot in 1 second.*

Watts and horsepower are directly related to the physical size of motors. Some motors are rated in kilovolt amperes (kVA) instead of horsepower. All transformers are rated in kVA.

The formula (volts × amperes = watts) shows that when the number of watts is constant, as the number of volts goes up, the number of amps goes down. For example, assume 1000 watts are required for a given load. If the power supply were 10 volts, it would take 100 amperes to produce 1000 watts of power. The wire size would have to be large to carry 100 amperes. With a 100-volt power supply, only 10 amperes are needed to produce 1000 watts. The wire size required would be much smaller. This is the reason large electric motors are designed to operate on high voltage.

Low amperes allow smaller wire to be used. Power lines are a good example of this. On the high-voltage (power line) side of a transformer, the wires are very small compared to those on the low-voltage side (load side).

Magnets and the Magnetic Circuits

The Bar Magnet

The bar magnet illustrated in Fig. 1.2 shows the invisible lines of force (flux) going out one pole and completing a magnetic circuit to the other pole. Lines of force will go through air, insulation, and nonmagnetic materials.

The bar magnet is at full magnetic strength when its molecules are in alignment, as seen in Fig. 1.3.

The Electromagnet

A piece of iron with a coil of wire wound around it makes a basic electromagnet (Fig. 1.4). Magnetic strength is controlled in an electromagnet by raising and lowering the amperes. Reversing the current flow will reverse its polarity.

The Magnetic Pole

A pole in the stator of a DC machine is a coil of wire wound around a piece of iron (called a pole shoe or pole iron), as shown in Fig. 1.5.

A pole is equal to 180 electrical degrees. One north pole and one south pole equal 360 electrical degrees. There are always pairs of poles.

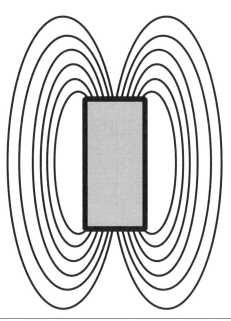

FIGURE 1.2   A bar magnet has lines of force that go from one pole to the other, through the air.

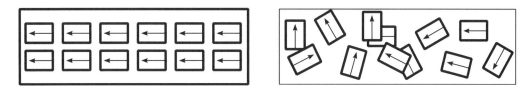

FIGURE 1.3   The magnet is fully magnetized if the molecules of the iron are aligned.

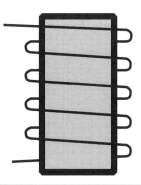

FIGURE 1.4   An electromagnet consists of an iron bar with a coil of wire wound around it.

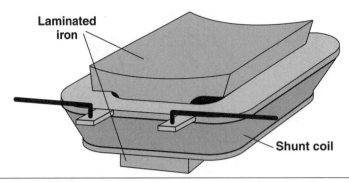

FIGURE 1.5   A field pole consists of laminated or solid iron and a coil of wire.

The bore (through the stator of the machine) is divided very precisely by the number of poles. The electrical degrees are equal to the mechanical degrees in a two-pole motor (as seen in Fig. 1.6). Half of the mechanical circle will contain 360 electrical degrees in a four-pole motor (Fig. 1.7).

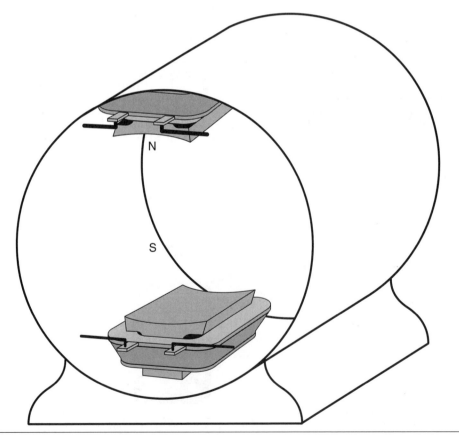

FIGURE 1.6   A two-pole stator.

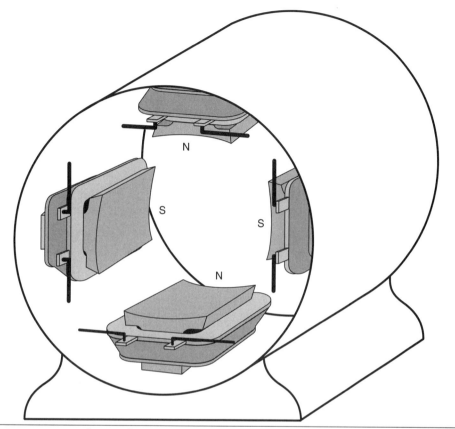

FIGURE 1.7   A four-pole stator.

Ampere Turns

Ampere turns is a term in a formula used in designing a pole. The strength of a pole is governed by the number of turns in its coil and the number of amperes flowing in them.

The shunt field of a DC machine has a large number of turns and a small number of amperes. Fewer turns and more amperes will do the same job but would be much more expensive to operate.

The series field in the DC machine is in a high-ampere circuit. It has very few turns, but its magnetic strength can be comparable to that of the shunt field.

The Magnetic Circuit

The circuitry of an electric motor sometimes seems very complicated. The armature of a DC motor is a good example. When the energized coils of the armature are isolated and displayed (Fig. 1.8), they take the same shape as the electromagnet shown in Fig. 1.4—a coil of wire around a piece of iron. Magnetic circuits in most motors are a variation of the basic electromagnet in Fig. 1.4.

A complete magnetic circuit (using a coil of wire as its power source) is shown in Fig. 1.9. The amount of current through the coil controls the number of lines of force (flux) in the magnetic loop.

When the magnetic loop is opened (Fig. 1.10), north and south poles are established on the ends of the iron. The direction of current flow through the coil determines the polarity at the ends of the iron. In a motor or generator, these poles are stationary and are part of the stator. The outer shell of the stator carries the lines of force from pole to pole.

If a bar magnet is placed in the opening (Fig. 1.11), the magnetic forces will cause torque. The bar magnet will try to align with the poles of the stator as shown in Fig. 1.12. (Unlike poles attract, and like poles repel.)

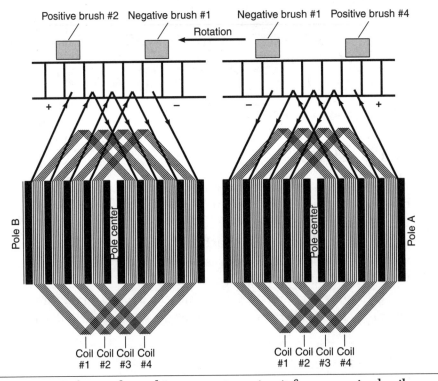

FIGURE 1.8 Poles are formed in an armature circuit from energized coils around idle coils.

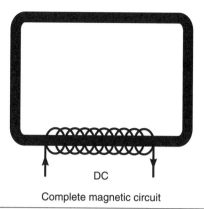

Complete magnetic circuit

FIGURE 1.9   An energized coil of wire around a loop of iron has a path for the magnetic flux it creates.

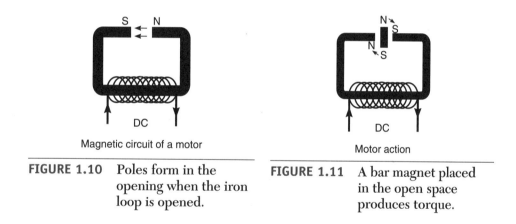

Magnetic circuit of a motor

FIGURE 1.10   Poles form in the opening when the iron loop is opened.

Motor action

FIGURE 1.11   A bar magnet placed in the open space produces torque.

Figure 1.13 shows the basic magnetic action of all electric motors. The poles of the bar magnet represent the magnetism developed in the armature of a motor.

Magnetic Saturation

The iron of a magnetic circuit has a limited capacity to carry lines of force. When this capacity is reached, it is called *magnetic saturation* or *fully magnetized*. When the iron's capacity is exceeded, it is called *oversaturation*. Figure 1.14 illustrates what happens to some of the lines of force when oversaturation occurs.

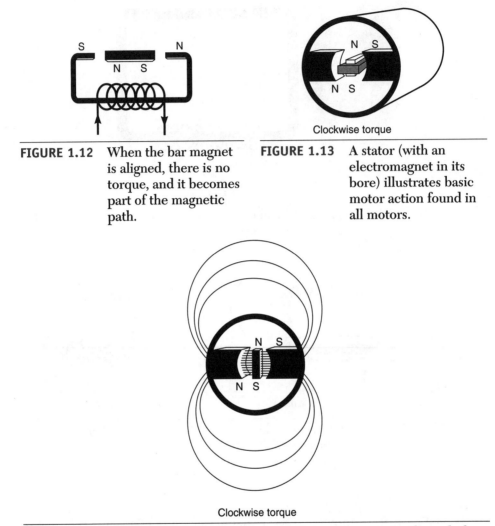

FIGURE 1.12 When the bar magnet is aligned, there is no torque, and it becomes part of the magnetic path.

Clockwise torque

FIGURE 1.13 A stator (with an electromagnet in its bore) illustrates basic motor action found in all motors.

Clockwise torque

FIGURE 1.14 Oversaturation causes magnetic lines of force to go through the air (when there isn't room for them in the iron).

Magnetic Balance

When electric motors and generators are designed, great care is taken to magnetically balance all the poles. The poles are placed an equal distance from each other around the stator. Each pole has the same number of turns of wire and produces the same amount of magnetism. The shell of the stator is the outer part of the magnetic circuit. The iron of the armature completes the inner part of the magnetic circuit.

Magnetic unbalance will cause bearing problems, loss of power, and internal heating in the armature.

Neutral Position in the Stator

The neutral spot is located an equal distance between the north and south poles. Magnetic neutral is 90 electrical degrees from each adjacent pole center. Interpoles (explained later in the chapter) are located in the stator's neutral position (Fig. 1.15). The correct brush setting will align the pole centers of the armature with the stator's neutral position on all DC machines.

Circuits of the DC Machine

A DC motor or machine has three basic circuits:

- Armature and interpole leads, A1 and A2

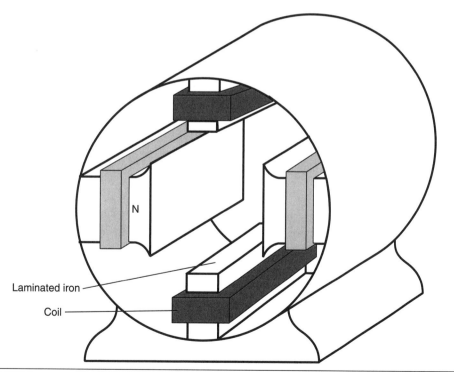

FIGURE 1.15   Interpoles in a two-pole machine—located at 90 electrical degrees—form the stator poles.

- Series field leads, S1 and S2
- Shunt field leads, F1 and F2

They are designed with various combinations and connections to suit the needs of a given load. A more in-depth explanation of each component will come later in this chapter.

As previously stated, the DC machine is interchangeable as a motor or a generator, so the word *machine* is used in this book when explanations apply to both motor and generator.

- The armature generates all the power as a generator. It creates all the torque as a motor. The armature and interpoles are a series circuit. The purpose of the interpoles is to improve brush commutation. They are connected either between A1 and the armature or between the armature and A2.
- The series field is connected in series with the armature and interpoles. Its purpose is to stabilize the output of the machine as the load changes.
- The shunt field provides magnetism for the armature. Lines of force or flux (produced by the shunt field) create power (VA) in the armature as a generator and create torque as a motor.

Rules for Generating Direct Current

When a conductor cuts or is cut by magnetic lines of force, a voltage is generated in it. The amount of voltage can be controlled by:

- The number of conductors in the armature
- The number of magnetic lines of force (or flux) from the stator
- The speed at which the armature conductors cut or are cut by the lines of force

The number of conductors is the same as the number of turns in the slot of an armature. Each turn of wire cuts lines of force and generates a given amount of voltage. The voltage generated in each turn is added to the next turn—this compares to flashlight batteries in series. The number of turns in the slots of the armature determines the basic voltage output of a generator.

The number of magnetic lines of force (being cut by the armature's conductors) controls its voltage output value. As the number of lines of force increases, the voltage value also increases in the conductors. The number of lines of force can be varied with a control that changes the number of amperes flowing in the stator fields. The output voltage will vary *with* the change in ampere flow through the stator fields. This control adjustment won't make a large change in the voltage output of a generator.

The speed at which the conductors cut the lines of force is determined by the generator's RPM. The recommended speed is on the generator's nameplate. The voltage output value will change *with* the speed change.

The Shunt Generator

The shunt generator has two circuits (Fig. 1.16), the armature and interpoles (A1 and A2) and the shunt field (F1 and F2).

Figure 1.17 shows the NEMA (National Electrical Manufacturers Association) standard connection for counterclockwise rotation facing the end opposite the shaft.

The Armature and Interpoles

Armature leads are identified as A1 and A2. The armature windings produce all of the generator's power output. The number of turns of wire in the slots

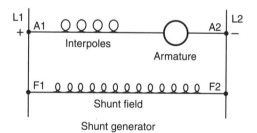

FIGURE 1.16   This schematic identifies the two circuits found in a shunt generator.

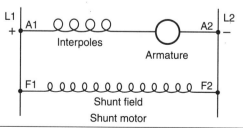

FIGURE 1.17   NEMA standard showing polarity and the lead number combinations for a shunt motor.

of the armature determines its voltage output value. The ampere rating of the generator determines the armature winding's wire size.

The armature circuitry consists of many coils that are connected to the commutator segments. The commutator segments control the direction of current flow in each coil of the armature. The basics of this circuitry and a more detailed explanation are provided in the section, "Operation of a DC Motor."

Interpoles are poles strategically placed in the stator to decrease brush arcing. They will be covered in depth later in this chapter.

The Shunt Field

The shunt field (Fig. 1.18) consists of coils of wire and laminated iron. Each coil and its iron make a pole. Fig. 1.19 shows a four-pole sketch. The coils are connected so that each is the opposite polarity from the one next to it. The number of north poles always equals the number of south poles. Figure 1.20 shows a four-pole stator.

The purpose of a generator's shunt field is to furnish magnetic lines of force (flux) for the armature conductors to cut. Power is produced as armature conductors cut the lines of force. The shunt field coils consist of hundreds of feet of wire. The total length of the wire in these coils controls the current. (Each wire size has a resistance value per 1000 feet.) The large quantity of wire in the shunt field circuit keeps its coils from overheating.

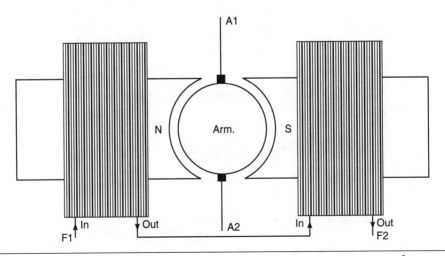

FIGURE 1.18   This sketch shows how pole-to-pole connections are made in a two-pole machine.

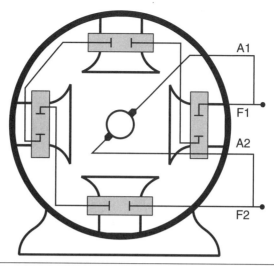

FIGURE 1.19   This sketch shows how pole-to-pole connections are made in a four-pole machine.

FIGURE 1.20   A four-pole shunt wound stator. *P&H MinePro Services.*

Operation of the Self-Excited Shunt Generator

The iron of the stator poles is still magnetized from its previous use. This is called *residual magnetism*. A few lines of force between poles are created by the residual magnetism. When the armature turns, its conductors cut the lines of force of the residual magnetism and create an *excitation voltage*.

The shunt field is connected in parallel with the armature circuit (Fig. 1.21), so the excitation voltage is applied to it. The excitation voltage creates a small amount of current flow in the coils of the shunt field. This small current increases the number of lines of force for the armature conductors to cut. The armature's voltage output increases as the current in the shunt field increases. Full voltage output can take several seconds. At this point, the shunt field pole iron is magnetically saturated. The generator is now ready to load.

As the load is applied, the voltage across the shunt field will drop slightly, lowering the shunt field amperes. This will cause the pole iron to have less than full saturation, and the output voltage will drop some. The shunt generator's full-load voltage will always be lower than its no-load voltage. Full-load voltage is the value given on the generator's nameplate.

Shunt Generator Control

Figure 1.22 depicts a shunt generator and its control. The control is used to regulate the generator's voltage downward a small amount.

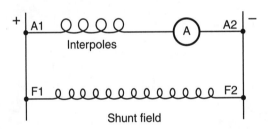

FIGURE 1.21   The shunt field completes a circuit for the power that the armature produces.

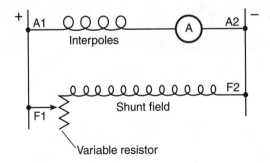

FIGURE 1.22   A rheostat that is used to adjust the voltage of the shunt field.

The resistance of the control lowers the number of shunt field amperes—which lowers the number of lines of force (flux) cut by the armature conductors, resulting in a lower voltage.

The shunt field can be excited separately with another power source. Separate excitation reduces the voltage drop common to the self-excited design.

The Series Generator

The series generator has two circuits that are connected in series with each other, the armature and interpoles (A1 and A2) and the series field (S1 and S2). Figure 1.23 shows a standard connection for counterclockwise rotation facing the end opposite the shaft.

All of the amperes produced by the armature pass through the series field. The series field coils are constructed with a few turns of wire that are large enough to carry the full ampere output of the generator. The large wire makes it a very low-resistance circuit compared to the shunt field circuit.

Operation of the Self-Excited Series Generator

The field pole iron has residual magnetism from the previous operation. The conductors of the armature cut lines of force of the residual magnetism and generate an excitation voltage. The output voltage will remain at the excitation voltage value until a load is applied. The excitation voltage will cause a small amount of current to flow through the load. This current goes through the series field, strengthening its magnetism and creating more lines of force.

When the armature conductors cut more lines of force, they generate higher voltage and amperes until the load demand is stabilized. The voltage output will remain at this value until more load is added. The voltage and ampere output will increase with the load increase until full power output of the generator is reached.

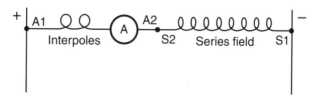

FIGURE 1.23   The schematic of a series generator.

At this time, the amperes in the series field create full magnetic saturation in the pole iron.

The series generator with no load will create only excitation voltage. The load regulates the voltage output. This characteristic limits its use.

The Compound Generator

The compound generator has three circuits (Fig. 1.24), the armature and interpoles (A1 and A2), the series field (S1 and S2), and the shunt field (F1 and F2).

Figure 1.25 shows the NEMA standard connection for counterclockwise rotation facing the end opposite the shaft.

The Armature and Interpoles

As shown earlier, the armature leads are identified as A1 and A2. The armature windings produce all of the generator's power output. The interpoles ensure good commutation.

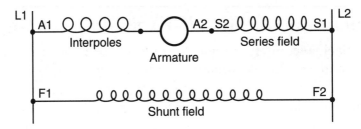

FIGURE 1.24   A compound generator, showing its three circuits.

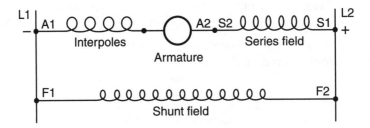

FIGURE 1.25   The NEMA standard connection for a compound generator, showing polarity and lead numbers.

The Series Field

The series field (S1 and S2) has coils that either are wound over the shunt field coils or are separately formed coils. If they are wound on top of the shunt field coils, they're separated with insulation. There is no internal connection between the series and shunt field coils. The series field coils can get hot because they are in a high-ampere circuit. They're located on the outside of the shunt coils, where they receive maximum cooling.

When the series field coils are separate, the coils are formed, insulated, and placed on the pole iron between the shunt field coils and the armature (Fig. 1.26).

The Shunt Field

As described earlier, the shunt field leads are identified as F1 and F2. The purpose of a generator's shunt field is to furnish magnetic lines of force (flux) for the armature conductors to cut. As the conductors of the armature cut the lines of force, power is produced.

Operation of the Self-Excited Compound Generator

Residual magnetism that is left in the pole iron from the previous operation creates a few lines of force. The armature conductors cut these lines of force and create an excitation voltage. The excitation voltage will cause a

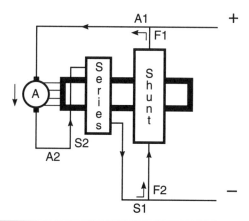

FIGURE 1.26 Current flow created by residual magnetism and the resulting lines of force.

small current to flow in the series and shunt fields (Fig. 1.26). This current increases the magnetic strength of the pole iron, creating more lines of force (which increase the voltage and current until full voltage output is reached). At this time, the pole iron has full magnetic saturation. This procedure can take several seconds. The shunt field furnishes nearly all the magnetism at no load.

As the load is applied, the load current produced by the armature goes through the series field, producing lines of force. These lines of force add to the shunt field's lines of force. Instead of the iron losing magnetism (as with the shunt generator), the series field maintains full magnetic power. The results are full field strength and no drop in output voltage. The added strength of the series field stabilizes the voltage as the load is increased.

The number of turns of wire in the series field directly affects the amount of stabilizing voltage it produces. The flat-compound generator has enough turns in the series field to raise the voltage as the load is applied. When this generator reaches its full rated load, the voltage is at nameplate value.

The over-compound design will have enough turns in the series field to raise the voltage to a higher value at full load than it had at no load. (The over-compound generator's output can be controlled downward to meet the requirement of the load.)

Compound Generator Control

Figure 1.27 shows a (basic) compound generator control. Its two main parts are a voltage control and a current control. (The control is designed for over-compound generators and controls the voltage downward.)

The design of these controls can vary from simple variable-resistor types to electronic ones. The more sophisticated type will react quickly to load changes and produce a very stable power output.

The voltage control varies the amperes of the shunt field. (Decreasing the shunt field amperes reduces the magnetic flux and lowers the output voltage.) This control doesn't make a large change in voltage.

The voltage control components are small compared to those of the current control because they control a small amount of current.

The current control diverts some of the amperes directly to the line, bypassing the series field. (The series field raises the voltage as the load increases.) The voltage increase is controlled by diverting some amperes.

The current control components are large because they control high amperes.

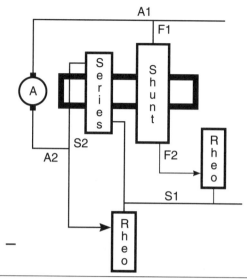

FIGURE 1.27   A compound generator with its control. The shunt field control is the voltage control, and the series field control is the current control.

The Cumulative Compound Connection (Generator)

The connection for the compound generator described earlier is called *cumulative compound*. If the series field is connected so that its polarity is the same as that of the shunt field, the connection is called *cumulative*. This connection adds the magnetic flux (lines of force) of the series field to that of the shunt field as the load increases. This tends to stabilize the voltage output.

The Differential Compound Connection (Generator)

When the series field polarity is opposite that of the shunt field, the connection is called *differential*. This connection (Fig. 1.28) has an effect that is opposite that of the cumulative connection.

The series field flux (lines of force) will cancel an equal amount of shunt field flux. This results in fewer lines of force for the armature conductors to cut, causing the voltage output of the generator to drop. If the load is decreased, the voltage will increase.

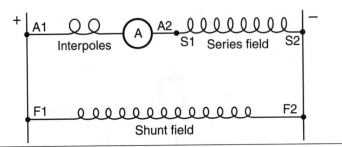

FIGURE 1.28   The differential connection, with the series field bucking the shunt field.

The differential connection in a generator is used to protect a load that can be damaged by high amperes. An example is a motor that is subject to stalling or regular overloading. The connection is used on large mining-equipment generators and all DC arc-welder generators.

Series Field Connections for a Motor and a Generator

The current flow through the armature of a motor is opposite that of a generator. A motor uses amperes from the line. A generator produces amperes.

The series field leads S1 and S2 have to be reversed when a motor is used as a generator, and vice versa. Figure 1.29 shows the NEMA standard connection for both applications. If a motor connection is used on a generator, the ampere flow through the series field makes the connection differential.

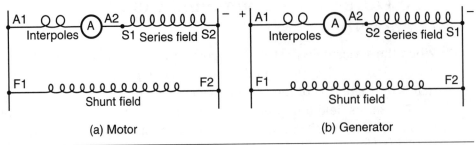

FIGURE 1.29   (a) The NEMA standard for a motor. (b) The NEMA standard for a generator.

Special-Purpose Machine Connections

Some special-purpose machines have an internal connection that connects the armature-interpole circuit with the series field circuit. They have one armature lead and one series field lead. Reversing these leads reverses the rotation and changes the connection from cumulative to differential.

The DC Arc Welder

The differential connection is used in a DC arc welder (Fig. 1.30). The components of a welder include:

- Exciter generator
- Shunt field
- Armature and interpoles
- Series field
- Reactor

The *exciter generator* is a separate DC source for the shunt field. (It's normally a small generator built into the machine.) It provides power for only the shunt field of the welder. Its power output is very constant and isn't affected by the welding load.

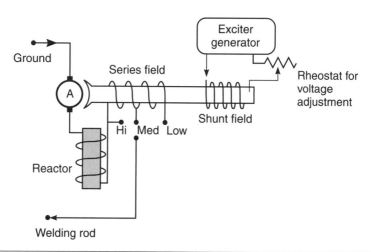

FIGURE 1.30 All DC welder generators have the same basic components, as shown here.

The *shunt field* is located on the same pole iron as the series field. It provides a constant source of magnetic lines of force for the conductors of the armature to cut. There is no electrical connection between the two circuits.

The welding load varies from the extremely high amperes of a short to the high amperes of an arc. It is basically a controlled short. For this reason, it's important to separately excite the shunt field so it's not affected by the demand of the welding load.

The shunt field requires a small number of amperes, so its control components are small. The output voltage of the welder is varied just a small amount with this control.

The *armature and interpole* conductors must be very large. Because of the low voltage and high ampere output, there are only a few turns of wire in the slots of the armature. (They're made of large flat wire.)

Interpole coils are connected in series with the armature. (Their coils are also made of large flat wire.) The interpole coil turns are spaced apart for maximum cooling.

The *series field* coils have very large conductors. The coil turns are spaced apart for maximum cooling (Fig. 1.31).

There are usually three or more taps connected to the series field. These taps select the number of amperes as needed and are the current control

FIGURE 1.31   Coils with the turns separated for maximum cooling.
P&H MinePro Services.

for the welder. The tap that includes all the series field turns will produce a maximum amount of canceling flux and will be the lowest ampere tap. The rest of the taps will include fewer of the series field turns. The highest ampere tap will include the least amount of series field turns.

The *reactor* coil consists of very large conductors. They are surrounded by iron (Fig. 1.30). The reactor's purpose is to oppose sudden changes in current flow.

Operation of the DC Welder Generator

When the welder is up to speed and ready to weld, the conductors of the armature are cutting the lines of force of the shunt field, creating full voltage. At this time, voltage between the welding rod and the metal to be welded is at its highest value.

When the welding rod strikes the metal, there is a short circuit and the ampere demand is very high. High amperes flow through the series field and create a canceling flux. The canceling flux lowers the voltage output of the generator, reducing the chance of the rod fusing to the iron. The reaction of the series field, however, isn't quick enough to prevent the fusing. (This is why a reactor is needed.)

Operation of the Welder's Reactor

Sudden high amperes are delayed (and lowered) as they go through the reactor. The delay is caused by the magnetizing of the reactor's iron.

This electrical delay allows the operator to start the welding arc without the rod fusing to the iron.

An arc has much higher resistance to current flow than when the rod is shorted to the metal. The higher resistance of the arc will cause a sudden drop in ampere flow. At this time, the reactor iron is highly magnetized. When the amperes drop suddenly, the magnetic field of the reactor collapses. As the magnetic field collapses, its lines of force cut the conductors of the reactor, creating a voltage. This voltage adds to the generator's voltage, maintaining the arc. Without the reactor, the arc would extinguish.

Without the reactor's effect, the rod would tend to stick or fuse to the metal. An arc would be hard to establish and maintain. The series field alone wouldn't react fast enough. The reactor is a very important part of the DC welder's operation.

Interpoles

The purpose of interpoles is to reduce brush arcing. An interpole is a coil of heavy wire around an iron pole piece. Interpoles are usually connected in series with each other. As a unit, they're connected between A1 and the armature or between the armature and A2. Their leads are usually connected internally to the brush holder (with no external accessible leads). Interpoles can be reversed internally to suit the application.

Some definite-purpose machines have an internal connection that connects the armature-interpole circuit with the series field circuit. In this case, there would be one armature lead and one series field lead.

The polarity of interpoles in a generator will be the same as the field poles ahead of them (in the direction of rotation) (Fig. 1.32).

The polarity of interpoles in a motor will be the same as the field poles behind them (in the direction of rotation) (Fig. 1.32).

There will be as many interpoles as there are field poles, or half as many. If there are half as many interpoles, each interpole coil will contain approximately twice as many turns.

Operation of Interpoles

The flux of two magnetic fields becomes distorted if they are close together. In a DC machine, the armature's magnetic field distorts the stator field's magnetic field. The amount of distortion varies with the change in armature load amperes. Distorted lines of force become a problem when they are forced into the neutral zone.

Armature conductors in the neutral zone are connected to commutator segments that are being switched by the brushes. The bar-to-bar voltage value should be only a portion of the line voltage. The conductors in this

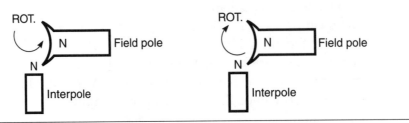

FIGURE 1.32   The polarity and direction of rotation for a generator (left) and for a motor (right).

position should not cut lines of force, and there should be no voltage generated in them.

The influence of the armature's magnetic field distorts the lines of force as shown in Fig. 1.33. Armature conductors in the neutral zone cut the distorted lines of force. This creates an unwanted voltage in the conductors. The unwanted voltage value adds to the line voltage value. The commutator's bar-to-bar voltage now becomes too high. The result is excessive arcing at the brushes.

The brushes can be shifted to reduce the arcing, but if the load changes, arcing will again become excessive. One of the reasons brushes arc excessively is that bar-to-bar voltage is too high for the insulation to contain it.

Interpoles are located in the neutral zone. The amperes of the armature go through the interpoles. This creates magnetic lines of force in the neutral zone (Fig. 1.34). These lines of force are cut by the armature conductors

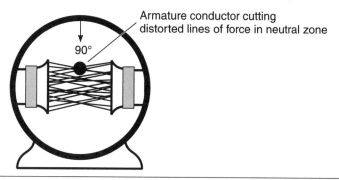

FIGURE 1.33   Magnetic lines of force become distorted by the armature's magnetism.

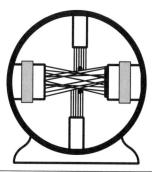

FIGURE 1.34   Magnetic lines of force from the interpoles are cut, creating a canceling voltage that neutralizes the unwanted voltage (that is produced by distorted lines of force).

generating a voltage in them. This voltage is the opposite polarity of the unwanted voltage described above and cancels it. The result is less arcing.

The polarity of the interpoles is very important. If the polarity is wrong, the voltage generated by them will add to the unwanted voltage instead of canceling it. The bar-to-bar voltage becomes very high, resulting in extremely excessive brush arcing.

The amount of field distortion is directly related to the number of amperes in the armature circuit. With the interpoles connected in series with the armature, the number of lines of force varies with the number of armature amperes. This automatically creates the right amount of canceling voltage needed to minimize brush arcing.

Interpoles must have exactly the right number of turns of wire. If the number of turns required involves a fraction of a turn, the next higher number of turns is used. In this case, nonmagnetic shims are placed under the iron of the interpole. They fine-tune the interpole's magnetic field to the exact strength needed.

Nonmagnetic shims are equivalent to increasing the air gap (the distance between the pole iron and the armature). Increasing the air gap weakens a pole's strength. Use of the nonmagnetic shims allows the pole strength to be weakened and still keeps the same air gap for all interpoles.

DC motor manufacturers custom-fit the interpole iron and shims. If they are disassembled, it's very important that the shims and pole iron be put back in their original location.

The Compensating Winding

The compensating winding (Fig. 1.35) is an extension of the interpoles. It is in series with the armature and is part of the interpole circuit. It brings the effect of the interpoles out to the center of the field pole iron. (This is necessary for motors with a large bore diameter and wide load variations.) When there is a sudden load change, the interpole's influence alone doesn't cover enough pole area to control the arcing.

The compensating winding usually consists of a single turn (a heavy copper bar) embedded in slots (cut in the face of the field pole iron).

Because of high current, the end turns and connections of the compensating winding extend beyond the shunt field coils. Any wire carrying high current that is located across a shunt coil's end turns will distort the coil's magnetic field. The result will be magnetic unbalance and brush arcing.

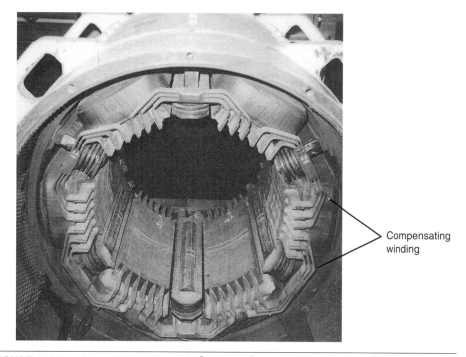

FIGURE 1.35   A compensating winding in a large DC machine. *P&H Mining Services.*

Operation of a DC Motor

DC generators and motors are interchangeable. Their armature circuits have a large wire and very low resistance and conduct all the ampere output as a generator.

When used as a motor, the low-resistance armature circuit is connected to line voltage. Because of its low resistance, it will demand destructively high amperes. The following text explains how amperes are controlled in a motor's armature circuit. It also compares an armature conductor generating voltage in a generator and the same conductor producing torque in a motor.

One Armature Conductor in a Generator as It Passes Two Poles

In a DC generator, the conductors of the armature cut lines of force provided by the stator fields and produce power. In Fig. 1.36, we can follow one conductor as it travels past the poles of a two-pole stator. The conductor's voltage value changes as its position changes.

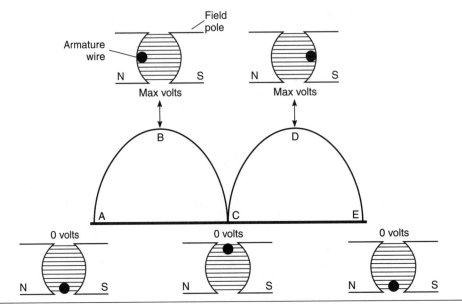

FIGURE 1.36   The amount of voltage (produced by a conductor) changes as it moves across a pole face.

Each stator pole is equal to 180 electrical degrees, making it possible to describe a location on a pole regardless of a motor's physical size. This particular magnetic circuit consists of two poles. Together they equal 360 electrical degrees.

Figure 1.36 illustrates the voltage values produced by a conductor as it cuts the lines of force of two poles. In position A, the conductor is not cutting lines of force because it is traveling in line with them. (This is the neutral position, where no voltage is produced.) As the conductor moves toward position B, it cuts more and more lines of force, increasing its voltage. When the conductor reaches position B, it will be cutting the maximum lines of force and creating maximum voltage. The conductor is now at the 90° spot of the pole.

As the conductor moves toward position C, it cuts fewer and fewer lines of force until it isn't cutting any lines of force (like position A).

As the conductor enters the south pole, it produces an opposite voltage. At this point, the commutator reverses the polarity of the conductor. The current continues to flow in the same direction as it did in position B (direct current). As the conductor moves toward position D, it cuts more and more lines of force, and its voltage increases. When it reaches position D, the voltage output again is maximum value. As the conductor moves toward position E, it cuts fewer and fewer lines of force until, like position A, position E voltage output is zero. The conductor has moved 360 electrical degrees.

One Armature Conductor in a Motor as It Passes Two Poles

Armature conductors in a DC motor create torque. Their position (as they move across a pole face) determines how much torque is produced. The ampere demand varies *with* the torque produced.

Figure 1.37 illustrates the torque value that one conductor contributes as it goes by two poles. In position A, no torque is produced because the conductor is traveling in line with the lines of force. When this happens, the conductor doesn't demand power from the line. As the conductor travels toward position B, the torque and ampere demand increase. At position B (90°), both torque and amperes are at their highest value. Torque and amperes decrease as the conductor moves to point C (180°).

As the conductor enters the opposite pole, the commutator reverses the direction of current flow. The conductor's torque remains in the same direction. The torque and power demand will increase until the conductor reaches point D (270°, or the 90° spot on this pole), where it reaches its

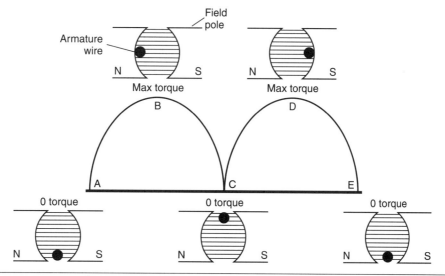

FIGURE 1.37   The amount of torque (produced by a conductor) changes as it moves across a pole face.

maximum value. Torque and amperes decrease as the conductor moves to point E. At point E, the conductor will have moved 360 electrical degrees.

Counter-voltage

As the armature of a motor turns, its conductors create torque and demand amperes from the power source. A motor's torque and ampere magnitude match a generator's voltage magnitude at the same location on the poles, as seen in Figs. 1.36 and 1.37.

Because the armature conductors are moving in a magnetic field, they are cutting lines of force and generating voltage. The voltage they generate is opposite the polarity of the power source. This is called *counter-voltage* (counter electromotive force, or counter emf).

The counter-voltage value depends on the speed of the armature and the strength of the stator field. The counter-voltage generated (as the motor's armature rotates) cancels an equal amount of line voltage. This lowers the amount of line voltage applied to the resistance of the armature circuit.

Counter-voltage keeps the amperes in the armature circuit at a safe value. The resistance of the armature circuit is so low that, without counter-voltage, the armature winding would be destroyed in a short time.

Counter-voltage Compared to Charging a Battery

Amperes controlled by counter-voltage can be compared to a battery being charged. (When a battery charger is attached to a dead battery, high amperes flow.) High amperes flow in a DC motor (Fig. 1.38a) when the armature is stationary and line voltage is applied.

The motor in this example is designed for 100 volts and 2 amperes. The armature has 5 ohms of resistance. One hundred volts divided by 5 ohms results in 20 amperes flowing through the armature at 0 RPM.

In Fig. 1.38b the battery voltage is up to 6 volts, or is "half-charged." Six volts in the battery cancel 6 volts of the charger, and the amperes drop. This compares to a motor turning at half-speed (Fig. 1.38b). At half-speed, the motor is generating 50 volts of counter-voltage. Fifty volts of counter-voltage cancels 50 volts from the line. Fifty volts divided by 5 ohms lowers the motor amperes to 10 amps.

Figure 1.38c shows the battery nearly charged with 11 volts. This cancels all but 1 volt of the charger, resulting in low amperes. This compares to the armature turning at its rated speed. In Fig. 1.38c, all but 10 volts are canceled

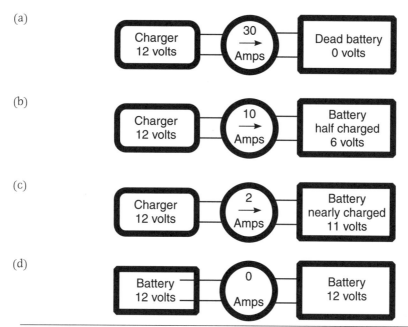

FIGURE 1.38   (a) A comparison is made between a dead battery getting charged and a loaded motor starting up. (b) A battery that is half-charged compared with a motor at half-speed. (c) A nearly charged battery compared to a motor running (at its rated speed). (d) Two fully charged batteries, connected as shown, will have no current flow between them.

by counter-voltage. This leaves 10 volts against 5 ohms of resistance in the armature, allowing 2 amperes to flow. This is the motor's full-load ampere rating.

If two batteries of equal charge are connected parallel (Fig. 1.38d), positive to positive and negative to negative, no amperes will flow between them. Twelve volts in one battery will cancel the 12 volts of the other.

If the motor's armature is somehow rotated fast enough, it generates counter-voltage equal to the line voltage. At this speed, there would be no amperes flowing in the armature circuit. If the armature is rotated even faster, the motor would become a generator.

Counter-voltage in a Motor Armature

The armature circuit has low resistance. When voltage is applied to it, a high inrush of amperes goes through the circuit. This gives the armature enough torque to rotate.

As the armature rotates, its conductors cut the lines of force of the shunt field. This action generates a counter-voltage in the conductors. The amperes of the armature are controlled by counter-voltage, the resistance of the wires, and the load.

As the armature gains speed, the counter-voltage increases and cancels more and more of the line voltage. As line voltage is canceled, amperes in the armature circuit go down. This is applying Ohm's Law logic: When volts are lowered, resistance doesn't change, and amperes decrease. At rated speed, only the amperes required by the load and the motor's internal losses are needed. A 500-volt motor develops full power with all but 10 to 15 line volts canceled by counter-voltage.

Ninety-five to ninety-eight percent of the armature circuit's potential current flow is canceled by counter-voltage. Without counter-voltage, high amperes and the resulting heat would destroy the armature. Unloaded, it would accelerate to a destructively high speed.

If the technician isn't aware of counter-voltage, testing the armature circuit will be very confusing. If the nameplate voltage and amperes are used to find the resistance of the armature circuit, the actual resistance is much lower.

Base Speed

Base speed is the number of RPM at full load (with full voltage applied to the armature and field circuits). At base speed, the motor is putting out nameplate horsepower and RPM.

Below Base Speed

Below base speed is accomplished by controlling the current through the armature, with full voltage applied to the shunt field. With the shunt field fully energized, the motor has constant torque (from zero to base speed). Consequently, the motor's horsepower output will increase as the speed increases.

This control is also used to start the motor.

Above Base Speed

The above base speed control is used after the motor has reached base speed, with full voltage applied to the armature circuit. As amperes are lowered in the shunt field, its magnetic field is weakened. A weaker magnetic field decreases the number of lines of force.

Fewer lines of force being cut by the armature conductors lowers the counter-voltage. Less counter-voltage allows more amperes to flow in the armature circuit. The motor accelerates to a speed above the original base speed and its speed is stabilized.

A weaker shunt field produces less torque, but with the increase in RPM, the motor still has the same horsepower. The control is also called *above base speed control torque control* or *field weakening control*.

The DC motor has constant torque below base speed and constant horsepower above base speed. These qualifications meet the demands of most loads.

Torque of the DC Motor

The DC motor has very good torque characteristics.

The commutator switches one circuit at a time in the armature slots. The slots contain one or more coils (circuits). As these coils are commutated, they control the magnetizing of the slot's iron teeth. As the magnetism in the iron of the armature changes from north to south, it does so one tooth at a time, resulting in a very smooth torque at all speeds.

The Shunt Motor

The shunt motor has two circuits: the armature and interpoles (A1 and A2) and the shunt field (F1 and F2) (Fig. 1.39). (A current control for starting the armature is also illustrated.)

Figure 1.40 shows the NEMA standard connection for counterclockwise rotation facing the end opposite the shaft.

The Armature and Interpoles

As shown earlier, the armature leads are identified as A1 and A2. The armature windings produce all of the motor's torque output. The interpoles ensure good commutation.

The Shunt Field

The shunt field leads are identified as F1 and F2. The purpose of a motor's shunt field is to furnish magnetic poles for the armature poles to attract or repel and create torque.

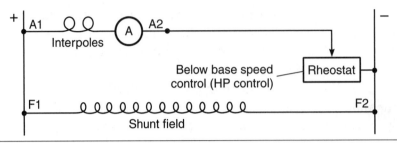

FIGURE 1.39   This schematic of a shunt motor shows the connections and the armature control.

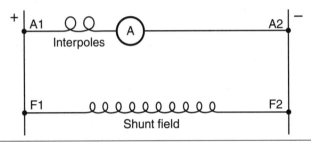

FIGURE 1.40   The NEMA standard shunt motor connection. Rotation is counterclockwise, facing the end opposite the shaft.

Operation of the Shunt DC Motor

When a shunt motor is started, the shunt field is energized with full voltage. An armature control puts resistance in series with the armature circuit, limiting the amperes to a nondestructive level. The control allows enough amperes through the armature to start the load. It then lowers the resistance until the motor accelerates to its selected speed.

If the shunt motor isn't loaded, it will accelerate to its no-load speed when full voltage is applied to its circuits. At this speed, counter-voltage allows only the amperes needed for the armature's losses. (This includes bearing friction, the cooling fan load, windage, and the armature circuit's copper losses.) A shunt motor's no-load RPM is higher than indicated on the nameplate.

As the load is applied to the shaft, the armature speed slows down. The armature conductors now cut fewer lines of force and create less counter-voltage. Less counter-voltage allows more amperes to flow in the armature circuit. The RPM will now stabilize at a lower speed. The load is then increased to the motor's rated horsepower. At full load, the motor will operate at nameplate amperes and RPM.

Starting a large DC motor with line voltage will cause excessively high amperes in the armature circuit and will damage the commutator.

If the shunt field disconnects or opens while the motor is running with no load, the armature will accelerate until it is destroyed. The armature will overheat if the motor is loaded and can't accelerate. Without the magnetism of the shunt field, the armature develops very little counter-voltage. Only the resistance of the armature windings will limit the amperes.

The Permanent Magnet DC Motor

A permanent magnet DC motor can withstand vibration and many other adverse conditions better than a motor with wound field coils.

The armature is the only electrical circuit in the permanent magnet DC motor. The stator has permanent magnets fastened to its shell. The permanent magnets aren't as thick as field coils, so the motor has a smaller diameter than the wound field coil motor.

Operation of the Permanent Magnet Motor

The permanent magnet motor operates much like a shunt motor. The magnets furnish torque and counter-voltage (like the coils of the shunt field).

The strength of the permanent magnets, however, can't be controlled. If the magnets lose strength, the motor will run faster than nameplate RPM. Loads that require more horsepower (as the speed increases) cause the armature to overheat and fail.

Restoring Magnetism

If a permanent magnet motor runs more than 10 percent above nameplate RPM (with no load), its magnets need to be remagnetized (*re-gaussed*). Magnetism in a weak magnet can be restored with a stronger magnet. An extremely strong magnet will bring a weak magnet to full strength almost instantly.

Electric motor repair centers use special equipment (a re-gaussing unit and an electromagnet) to re-gausse weak magnets. The re-gaussing unit contains a large number of DC capacitors. When fully charged these capacitors can—for an instant—discharge a very high DC current.

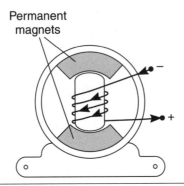

FIGURE 1.41   An electromagnet made for re-gaussing weak permanent magnet motors.

An electromagnet is designed to closely fit the bore of the permanent magnet motor. It's placed in the motor's bore, as illustrated in Fig. 1.41. The re-gaussing unit's capacitors are discharged through the electromagnet's coil—once or twice—creating very strong magnetism. The motor's magnets are almost instantly re-gaussed. The motor is then assembled and test run.

Large permanent magnet motors have magnetizing field coils wrapped around the permanent magnets. These field coils demagnetize and remagnetize the permanent magnets. (It's very hard to remove the armature unless the permanent magnets have been demagnetized.) After the motor has been repaired and the motor reassembled, DC is applied to the magnetizing field until the permanent magnets are remagnetized. The permanent magnets have full strength when the motor runs less than 10 percent above nameplate RPM with no load. (The magnetizing field leads are separated and insulated when they aren't in use.)

The stator's permanent magnets will lose power if the armature is left out of the stator for an extended time. A closefitting rotor, a piece of iron, or another armature should be inserted into the bore. This completes the magnetic path. It's similar to the keeper that is used on horseshoe magnets. Magnetism can also be weakened if the motor is reversed (plugged) while running. Another suspected cause is rough handling.

The Series Motor

The series motor has two circuits, the armature and interpoles, A1 and A2, and the series field, S1 and S2 (Fig. 1.42).

Figure 1.43 shows the NEMA standard connection for counterclockwise rotation facing the end opposite the shaft.

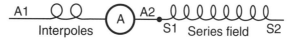

FIGURE 1.42   The circuits of a series motor.

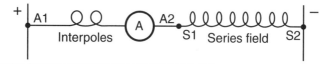

FIGURE 1.43   The NEMA standard schematic of a series motor. Rotation is counterclockwise facing the end opposite the shaft.

As shown in Fig. 1.43 the two circuits are connected in series with each other. All of the armature amperes pass through the series field. This makes it necessary to construct the series field coils with wire large enough to carry the full amperes of the motor.

The large wire used in the series field coils makes it a very low-resistance circuit. Amperes through the series field are limited by the counter-voltage developed in the armature. A series field (S1 and S2) should never have the motor's rated DC voltage applied to it.

Operation of the Series Motor

The series motor has very high starting torque. The ampere demand of the armature is very high (at 0 RPM). The same high amperes flow through the series field, resulting in very strong starting torque.

As the armature accelerates, it develops counter-voltage. The counter-voltage lowers the amperes through the armature and the series field. Fewer amperes through the series field lowers the number of lines of force. This lowers the amount of counter-voltage and torque.

The counter-voltage never becomes high enough to control the speed of the series motor, as it does with the shunt motor. Only the load will keep this motor from destroying itself because of high RPM. (Technically, it has no top speed.) A large series motor must have "no-load protection" in its control. The load is usually direct drive, so this motor must always be loaded.

The starter for an automobile engine is a series motor. Although it's small, it has very high starting torque.

Power tools also have series motors. Gear and bearing friction and a cooling fan keep them from developing excessive RPM.

The Compound DC Motor

The compound DC motor has three circuits (Fig. 1.44), the armature and interpoles (A1 and A2), the shunt field (F1 and F2), and the series field (S1 and S2). Figure 1.45 shows the NEMA standard connection for counterclockwise rotation facing the end opposite the shaft.

The Armature and Interpoles

As shown earlier, the armature and interpole leads are identified as A1 and A2. The armature windings produce all of the motor's torque output. The interpoles ensure good commutation.

The Shunt Field

As shown earlier, the shunt field leads are identified as F1 and F2. The purpose of the shunt field is to furnish magnetic lines of force (flux). The armature conductors interact with the shunt field flux to produce torque and counter-voltage.

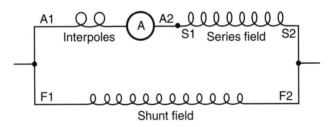

FIGURE 1.44   The compound motor with the armature, interpoles, series field, and the shunt field.

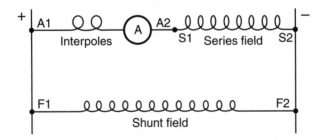

FIGURE 1.45   The NEMA standard connection for the compound motor. Rotation is counterclockwise, facing the end opposite the shaft.

The Series Field

As described earlier, the series field (S1 and S2) is connected in series with A1 and A2. The armature current goes through the series field, adding magnetic strength to the pole iron. This stabilizes the motor's speed when the load varies.

Operation of the Compound Motor Connected Cumulative Compound

The shunt field is connected in parallel with the armature and series field circuit. Unloaded, the motor starts and accelerates to a speed that is determined by the shunt field. (The shunt field produces enough counter-voltage to stabilize the speed.)

As the load is applied, the amperes increase in the armature. The armature current goes through the series field, producing lines of force. These lines of force are added to the shunt field lines of force, resulting in an increase in field strength and torque. The added strength of the series field stabilizes the RPM as the load is applied. Unlike the shunt motor, this motor doesn't slow as the load increases.

The number of turns of wire in the series field directly affects the amount of stabilizing done.

The flat compound design has enough turns in the series field to raise the RPM as the load is applied. When rated load is reached, the RPM stabilizes at nameplate value. The over-compound design has enough turns in the series field to raise the RPM to a higher value at full load than at no load. This is the more popular design. The motor's control gives it a wide speed range.

The Stabilized Shunt Motor

The stabilized shunt design has very few turns in the series field. The series field is connected differential compound. As the load increases, armature and series field current increases.

The resulting series field magnetism cancels some of the magnetism of the shunt field. This lowers the counter-voltage very quickly and allows more amperes to flow in the armature. The motor will react quickly to sudden load changes. Some efficiency is lost with this design.

Speed Control for a Compound DC Motor

The motor's nameplate shows the base speed of the motor. Figure 1.46 shows a basic compound motor control. Controls can range from hand-cranked rheostats to electronic circuitry. Regardless of their complexity, they will control the components of the motor as illustrated in Fig. 1.46.

Armature Control for Starting and for Below Base Speed

Full voltage should be applied to the shunt field when the armature control is used. Full voltage to the shunt field will give the motor constant torque (from zero to base speed). The armature control allows enough amperes through the armature to give it breakaway torque. The resistance is lowered gradually as the motor starts to accelerate. The motor accelerates to the RPM required by the load. Full voltage is applied to the armature at base speed.

Shunt Field Control for Starting and for Above Base Speed

The shunt field control (Fig. 1.46) applies full voltage to the shunt field when the motor starts. The motor accelerates until the RPM is stabilized at base speed. At this RPM the lines of force furnished by the shunt field are at maximum. The amperes flowing in the armature are limited by counter-voltage to a value needed for the load and the motor's internal losses.

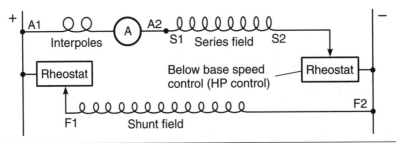

FIGURE 1.46   The basic control for a compound DC motor. The armature rheostat limits the current for starting and controls the below base speed. It is sometimes called the horsepower control. The shunt field control is the above base speed control and is sometimes called the torque control.

The shunt field control (sometimes called above base speed, field weakening, or torque control) can now be used. It adds resistance to the shunt field, lowering the amperes. This will reduce the shunt field's magnetic lines of force or flux. Less magnetism from the shunt field lowers the counter-voltage and allows more amperes to flow in the armature. The result is a higher base speed. The torque is reduced, but increased RPM gives the motor constant horsepower above base speed.

All controls have a field loss feature that shuts the motor off if the shunt field circuit opens. If the shunt field disconnects or opens while the motor is running with no load, the armature will accelerate until it is destroyed. If the motor is loaded and can't accelerate, the armature will burn out. Without the magnetism of the shunt field, the armature develops very little counter-voltage. Only the resistance of the armature windings limits the amperes.

The control components used to control the shunt field are small because of the small number of amperes the shunt field requires.

The shunt field can be excited separately with a different power source. With separate excitation, the shunt field's supply voltage isn't affected by the high amperes of the armature circuit.

The Differential Connection (Motor)

The connection for the compound motor (described earlier) has the series field connected so that its polarity is the same as that of the shunt field. This connection is called *cumulative*. It adds the magnetic flux (lines of force) of the series field to that of the shunt field, enabling the motor to maintain stable RPM as the load changes.

When the series field polarity is opposite that of the shunt field, the connection is called *differential*. With the differential connection (Fig. 1.47), the series field has an effect that is opposite that of the cumulative connection (instead of adding to the shunt field magnetism, the series field reduces it).

As the load increases, the series field amperes increase. If connected to the opposite polarity as the shunt field, the series field's flux (lines of force) cancels an equal amount of shunt field flux. This reduces the lines of force being cut by the armature conductors, lowering the counter-voltage.

Lower counter-voltage allows the amperes through the armature to increase. Higher armature amperes will accelerate the motor to a higher speed. Higher speed increases the counter-voltage, lowering amperes in the armature and series field circuit.

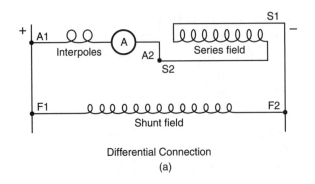

Differential Connection
(a)

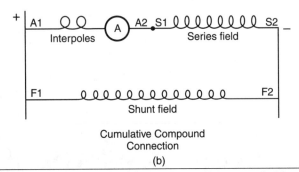

Cumulative Compound
Connection
(b)

FIGURE 1.47   In the differential connection (a) the series field is connected opposite that of the shunt field. The cumulative connection (b) is used unless the series field is designed for the differential connection.

The (fully energized) shunt field flux quickly brings the armature speed back down. When the load demand again slows the armature, the cycle starts again. The result is wide swings in speed.

Armature Components

- Commutator
- Commutator film
- Armature windings
- Equalizer shunts
- Laminated iron

Commutator

The commutator is made of copper segments that are separated and insulated with mica. The commutator segments function as one side of a switch, and the brushes function as the other. Together they control the direction of the current flow (similar to a diode). The commutator's function is to control the polarity of the coils connected to its segments.

As the armature turns, two commutator segments (and the coil attached to them) have current flow in one direction. The coil is now part of pole A (Fig. 1.48). When the coil's segments go under a brush, they short together, causing line current to stop flowing in the coil.

When the coil's segments come out from under the brush, the current is reversed, and the coil is now part of pole B—the opposite polarity pole in

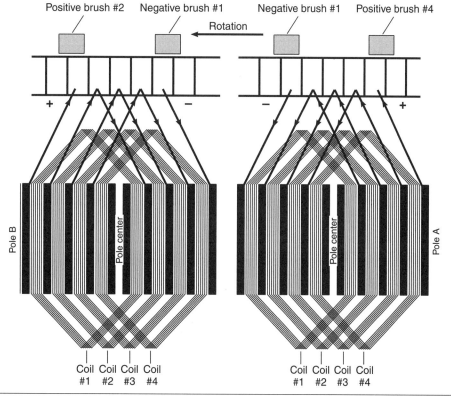

FIGURE 1.48   The commutator bars and armature coils flattened to illustrate how a single coil's polarity is switched to form poles that are similar to stator poles.

the direction of rotation. No matter how fast the armature turns, the poles in its iron will remain in the same position (relative to the stator poles). The commutator and brushes are high-maintenance components because of friction and electrical wear. (Electrical wear is caused by constant switching.)

Commutator Film

Good film is a very important part of the commutator function. It conducts current and provides lubrication between the brushes and the copper segments. The film shouldn't be removed unless the commutator is reworked.

It takes from several hours to several days to reestablish the film. During this time, the brushes and the commutator are wearing excessively.

The film has some resistance, and removing it will increase the voltage output of a generator and increase the speed of a motor. Slightly better performance, however, isn't worth the amount of increased wear to both brushes and commutator.

The film is roughly 1/1000 the thickness of a human hair. Elements of the film include carbon, graphite, copper oxide, and water vapor. Carbon, graphite, and copper oxide are current-conducting elements of the film. Water vapor and graphite provide lubrication. The brushes and commutator need a good balance of these elements for maximum life.

Lack of oxygen lowers the copper oxide element of the film, reducing its current-carrying ability. This occurs in totally enclosed and explosion-proof machines.

Dry weather reduces the water vapor content of the film, resulting in increased wear.

There is no "best color" for the film, and it can vary from machine to machine. The color can vary from tan to black and still do a good job.

A color difference in several adjacent bars (Fig. 1.49), with this pattern repeated all around the commutator, doesn't indicate a problem. It means there is more than one circuit (or coil) per slot. As the armature rotates, the circuits in this slot are commutated at a slightly different position (relative to the field poles). If the bars commutate in different positions, each bar commutates a different current value. (This causes each bar to have a different color.)

The color pattern should be the same around the commutator. A difference in the color pattern may indicate a problem.

FIGURE 1.49   Slot bar marking caused by a number of circuits in the armature slots. *Helwig Carbon Products, Inc.*

If there is bar etching, it can be reduced by moving the brush holder slightly less than one segment. Move the brush holder *against* rotation if it's a motor and *with* rotation if it's a generator.

Armature Windings

Armature windings are made with large wire. Most of the amperes flow through the armature in a motor. All of the output amperes flow through the armature in a generator.

Lap and wave are the two basic connections for the armature windings. The lap connection has as many brushes as there are poles. The wave connection has half as many brushes as there are poles.

Whatever the connection, the poles are precisely spaced and balanced magnetically. Most windings have the same number of turns per slot. If they don't, the total turns per pole will be the same. This fact makes it possible to use the comparison test when troubleshooting.

Equalizer Shunts

Equalizer shunts (back connections) are used to balance the effects of voltage unbalance in the armature windings. They are placed in the commutator's wire slots or risers. They join segments located under like pole brushes.

Unbalanced voltage will occur between a brush and its two adjacent brushes (Fig. 1.50). The resulting current will flow in the equalizer shunt wire instead of the brushes and commutator. At full load, the commutator is already carrying its rated amperes.

Without equalizer shunts, the amperes caused by unbalanced voltage adds to the full-load amperes. The resulting commutator damage is the same as if some of the commutator bars were overloaded. High current in these bars will cause them to discolor and erode. Bar erosion will eventually cause a flat spot on the commutator.

Laminated Iron

Laminated iron is the magnetic path for the armature. The laminations are thin because of a high-frequency magnetic polarity change in the iron teeth. The teeth in a high-speed DC armature will have a polarity change well over 100 cycles per second.

A thin coat of insulation separates the laminations. If the laminations are shorted together, eddy current is generated. This current will develop a hot spot in the laminations that will damage the adjacent slot insulation.

Repair centers always core test and, if necessary, re-insulate the laminations before rewinding an armature.

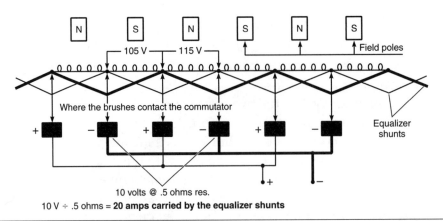

FIGURE 1.50   Equalizer shunts connected to like polarity commutator bars.

Chapter 1 Review

1. Voltage (page 2)
   a. causes current to flow.
   b. must have heavy wire for high voltage.
   c. must have current flow to exist.

2. Voltage value determines the thickness of the insulation (page 2).
   T___ F___

3. Heavy copper wire has no resistance (page 2). T___ F___

4. Resistance controls the voltage (page 2). T___ F___

5. Name two forms of resistance (page 2).

6. Ampere flow is controlled with voltage value and resistance value
   (pages 2–3). T___ F___

7. Two factors for determining wire size are: the number of amperes and
   the cooling capability of the device (page 3). T___ F___

8. Amperes can be controlled (up or down) without changing the voltage
   or resistance (pages 2–3). T___ F___

9. As its heat increases, copper will have (page 5)
   a. more resistance.
   b. less resistance.

10. Ohm's Law proves that, as voltage increases, amperes go up (if the
    resistance stays the same) (page 5). T___ F___

11. A watt is a measurement of power consumed (page 5). T___ F___

12. As watts go up, amperes go down (page 6). T___ F___

13. The number of watts determines the required wire size (page 6).
    T___ F___

14. Magnetic lines of force go through air and insulation (page 6). T___ F___

15. Strength of an electromagnet is controlled by the number of amperes
    flowing through its coils (page 6). T___ F___

16. Polarity of an electromagnet is controlled by the number of amperes
    flowing through its coils (page 6). T___ F___

17. One north pole and one south pole in a six-pole motor equal _____
    electrical degrees (pages 6–8).

18. A six-pole motor has _____ mechanical degrees (page 8).

19. A two-pole motor has _____ electrical degrees (page 8).

20. The strength of an electromagnet is controlled by the number of amperes that flow through its coils, and the number of turns (page 9). T___ F___

21. The magnetic strength of a pole with 10 turns, and 10 amps flowing through it, is the same as a pole with 20 turns, and 5 amps flowing through it (page 9). T___ F___

22. A complete magnetic circuit is vital to the operation of a motor or generator (pages 9–10). T___ F___

23. All motor and generator poles resemble an electromagnet (page 10). T___ F___

24. Magnetic saturation is reached when the iron in a motor or generator is fully magnetized (page 11). T___ F___

25. Magnetic balance requires that all poles have the same strength, and that they are equally spaced in a machine (page 12). T___ F___

26. The neutral spot in all stators is located equal distance from adjacent pole centers (page 13). T___ F___

27. When the armature poles align with the stator poles, the brush setting is correct (page 13). T___ F___

28. Name the three basic circuits of the DC machine (pages 13–14).

29. The armature creates _____ in a generator and _____ in a motor (page 14).

30. The voltage output of a generator is controlled by the number of conductors in its armature circuit, the number of lines of force, and the speed at which the armature conductors cut the lines of force (page 14). T___ F___

31. Draw the connection schematic of a shunt generator (page 15).

32. Amperes determine the size of wire used in an armature winding (page 16). T___ F___

33. The length of the wire in the coils of the shunt field controls the amperes (page 16). T___ F___

34. What is the purpose of residual magnetism (pages 17–18)?

35. Lowering the amperes in the shunt field raises the voltage output (page 19). T___ F___

36. The series field coils are made of heavy wire because all of the ampere output goes through this circuit (page 19). T____ F____

37. A series generator with no load will produce full voltage (page 19). T____ F____

38. Draw the NEMA standard connection for a compound generator (page 20).

39. The purpose of the series field in a compound generator is to stabilize the voltage output (page 22). T____ F____

40. The power output of an over-compound generator is controlled downward by a control (page 22). T____ F____

41. Diverting some series field amperes will reduce the series field's ability to increase the generator's output (page 22). T____ F____

42. The cumulative compound connection in the compound generator reduces the voltage output when the load increases (page 22). T____ F____

43. The differential compound connection in the compound generator reduces the voltage output when the load increases (page 23). T____ F____

44. The series field is connected the same in a motor as in a generator (page 24). T____ F____

45. An arc welder is connected differential (page 25). T____ F____

46. The shunt field of a welder is separately excited (page 26). T____ F____

47. Selecting a tap that adds more turns to the series field will increase the amperes (pages 26–27). T____ F____

48. A reactor in the welder's circuit bucks sudden changes and makes welding easier (page 27). T____ F____

49. Interpoles are connected in the armature circuit (A1 and A2) (page 27). T____ F____

50. The polarity of interpoles in a generator will be the same as the field poles behind them (in the direction of rotation) (page 28). T____ F____

51. The polarity of interpoles in a motor will be the same as the field poles ahead of them (in the direction of rotation) (page 28). T____ F____

52. Interpoles create a canceling voltage that decreases brush arcing (pages 29–30). T____ F____

53. What happens when the interpoles are connected wrong (page 30)?

54. If interpoles are removed from a stator, it is vital that they (and their shims) be reassembled exactly as they were originally (page 30). T___ F___

55. The compensating winding is necessary in machines that have large bores, and that have sudden load swings (page 30). T___ F___

56. What controls the amperes in a motor that is running at full speed (page 34)?

57. Why is there no ampere flow between batteries that have equal voltage (page 35)?

58. Counter-voltage controls amperes in the armature circuit of a motor (page 36). T___ F___

59. If a motor's armature is rotated fast enough, it becomes a generator (page 35). T___ F___

60. Define base speed (page 36).

61. Below base speed is achieved by reducing the armature amperes, and applying full voltage to the shunt field (page 36). T___ F___

62. Above base speed is achieved by applying full voltage to the armature, and reducing the strength of the shunt field (page 36). T___ F___

63. A motor's torque is reduced when the shunt field is weakened (page 37). T___ F___

64. A motor's horsepower remains the same when the shunt field is weakened and the RPM increase (page 37). T___ F___

65. RPM of a shunt motor are higher with no load than it is with full load (page 38). T___ F___

66. Large DC motors will suffer commutator damage if started across the line (page 39). T___ F___

67. An open shunt field will cause a motor (with no load) to stop (page 39). T___ F___

68. The speed of the permanent magnet motor will increase if the magnets become weak (page 39). T___ F___

69. The permanent magnet's strength is acceptable if the motor runs 20% higher than its nameplate RPM (page 39). T___ F___

70. A permanent magnet's strength can be restored almost instantly with a re-gaussing unit (page 40). T___ F___

71. A series motor's field can have full voltage applied to it (like a shunt field) (page 41). T___ F___

72. The series motor has low starting torque (page 41). T___ F___

73. The series motor's torque decreases as its RPM increases (page 41). T___ F___

74. What is the series motor's top speed (page 41)?

75. Draw the NEMA standard connection for a compound DC motor (page 42).

76. The no-load speed of the compound motor is determined by the shunt field (page 43). T___ F___

77. The stabilized shunt motor is connected differential to make the motor react quickly to load changes (page 43). T___ F___

78. The stabilized shunt makes a motor more efficient (page 43). T___ F___

79. Starting a motor and below base speed both require limiting the current in the armature circuit (page 44). T___ F___

80. If the shunt field amperes are limited when the motor starts, the amperes of the armature become excessively high, and the torque is lowered (page 44). T___ F___

81. Connecting a cumulative compound connected motor differential can cause erratic speed swings when it's loaded (pages 45–46). T___ F___

82. The commutator is one side of a sliding switch (that acts as a diode) (page 47). T___ F___

83. Commutator film should be cleaned off at regular intervals to lengthen the life of the commutator (page 48). T___ F___

84. Armature windings carry most, and sometimes all, of a DC machine's amperes (page 49). T___ F___

85. Equalizer shunts connect commutator segments that are located under like pole brushes (page 50). T___ F___

86. Laminated iron reduces eddy current (page 50). T___ F___

Chapter 2

Troubleshooting and Testing the DC Machine

Armature Problems

A high percentage of a DC machine's problems occur in the commutator and brushes. Following are some of the test procedures, problems, their causes, and solutions.

Burned Commutator Bars

Burned commutator bars indicate a serious problem, and the machine will need further testing. If two spots appear at 180 electrical degrees from each other (two adjacent brush holders apart on the commutator of most armatures), the armature may have shorted coils. An armature with shorted coils should be rewound.

In some cases, the shorted coil can be cut open and the commutator segments attached to it can be shorted together. (This should be done only in an emergency.)

Dirty Commutator

The mica between the segments has to withstand low voltage, high heat, and extreme mechanical pressure. Because of the heat, this mica has no bonding or sealing agents. It has a linear grain that is flawed with voids.

Thin liquids should not be used to clean a commutator. If lacquer thinner or a similarly thin liquid is used, it will wash brush dust into the voids and seriously degrade the mica's insulating ability.

Wipe or scrape the commutator to clean it. A cloth slightly dampened with solvent can be used. A sign of shorted mica is bright green arc streamers partway around the commutator.

Dead Spots from an Open Armature Coil

The indication of an open armature coil is a single burned spot on the commutator involving two bars. The leading bar (in the direction of rotation) will have material arced away because it breaks the circuit. The following bar may be discolored (from the circuit breaking arc) but will not be damaged.

An open coil in a generator armature will cause a fluctuating voltage output, and an open coil in a motor's armature may cause dead spots. The motor won't start unless the armature is rotated slightly.

Opens are often found where wires are soldered to the commutator segments. This can be repaired by soldering. (Sometimes this problem can be repaired in place.)

Open Equalizer

When two spots occur that are equal to the spacing of like pole brushes, an equalizer (explained under "Equalizer Shunts" in Chapter 1) is open. They can be resoldered and, in some machines, repaired in place.

Most brush companies have colored pictures of commutation problems, which are very helpful to their customers. The preceding descriptions of shorts and opens aren't usually illustrated.

Stalled Motor

A motor that is damaged (from being stalled) will develop discolored commutator segments—spaced the same as the brush locations. Even when cleaned, the excessively heated commutator segments will again become discolored as the motor operates. If these segments wear and create flat spots, extensive repair is needed.

Cleaning-action brushes can be installed to reduce this type of discoloring on smaller motors. They shorten commutator life and should be used sparingly. Under maximum load, they will arc more than normal brushes because they don't conduct current as well.

Individual segments can be replaced in large commutators. In smaller commutators the whole commutator must be replaced.

This same color pattern will appear if the load stops the armature in the same spot every time. If possible, the load should be adjusted to avoid this problem.

Control and Load Problems

Any quick speed change will shorten the life of the commutator and the brushes. (Large motors must have the armature amperes reduced when starting.) A sudden speed increase will damage the commutator as much as across-the-line starting. Sudden load changes will also cause commutator damage.

In some cases, the interpoles won't have time to react, causing the brushes to arc excessively for a short time. (Machines with a large bore have compensating windings to reduce this problem, as explained previously under "Interpoles" in Chapter 1.)

Brushes

Brushes are classified in the following groups: carbon-graphite, electrographitics, graphites, and metal graphites. Each has its own current-carrying, wear, and filming characteristics. They can be designed for specific requirements such as low voltage/high current, dry conditions, cleaning action for overfilming, and high speed.

Brush Problems

The carbon brush used in a DC machine has a difficult job:

- It must rub on the commutator hard enough to conduct amperes.
- It has to tolerate high heat from friction, ampere flow, and arcing.
- It should cause minimal wear of the commutator material.

A properly designed brush accomplishes all of the above. Original equipment brushes don't always work. Ambient conditions sometimes require a different mix of brush elements. Aftermarket brush companies specialize in solving these problems.

Following are some of the brush and commutator problems and solutions for the DC machine.

Conduction Zone

There is a small area of high current on the face of the brush called the conduction zone (Fig. 2.1). The zone current is part load current and part circulating current.

Carbon conducts better as its temperature goes up. This contributes to the current concentration in a small area. The conduction zone is roughly 10 percent of the total brush face area. (There can be more than one conduction zone on a brush face.)

A circulating current is caused by a brush shorting two commutator segments (connected to an energized armature coil). When the brush shorts the coil's two segments, the line current stops, and the magnetism of the coil collapses. This action creates a voltage in the coil.

The brush has joined the segments at the ends of the coil, creating a closed loop circuit. This circuit has low resistance, so a pulse of high current flows in it (Fig. 2.2).

The current flows in the opposite direction from what it did when the coil was part of the previous pole. It flows in the same direction as it will when the coil becomes part of the next pole.

The commutator will wear evenly if the conduction zone moves back and forth under the brush. (The zone is moving as it should if sparking at the trailing edge of the brush moves back and forth.) If it doesn't move, or if the circulating current in the zone is too high, the commutator will be damaged.

Two commutator bars are conducting the circulating amperes of the zone and the load amperes. If the circulating current causes excess heat, it's

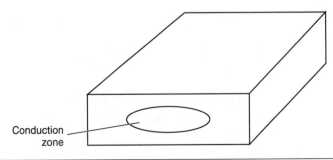

Conduction zone

FIGURE 2.1   The conduction zone of a brush, where most of the current flows.

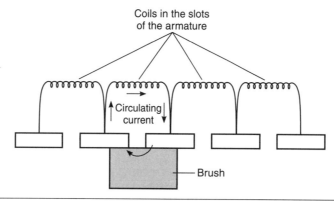

FIGURE 2.2 A discharge current flows in the closed loop when the brush shorts out the two bars connected to the coils.

too high and must be reduced. Higher-resistance brushes will reduce the circulating current; however, they also lower the operating amperes.

Brushes that are split into two or more equal parts help control circulating current. The resistance between the brush segments lowers the circulating current without affecting the operating amperes. A split brush also enhances the movement of the conduction zone. The brush sections divide the zone area, leaving each brush section to carry only part of the zone.

Spiral Groove

A machine is sometimes severely overloaded during its duty cycle, causing commutator damage. A spiral groove cut into the segments of the commutator (Fig. 2.3) will reduce or eliminate this damage and force the conduction zone to move.

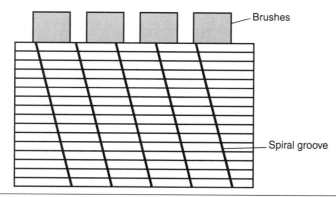

FIGURE 2.3 A spiral groove cut in the bars of a commutator to improve commutation.

The groove should be about twice as wide as the undercut slot of the mica, but not quite as deep. At this depth, the groove will be removed the next time the commutator is reworked on a lathe. (It's easier to cut a new spiral than to deepen it.)

The groove should sweep across the face of a brush at least twice per revolution. Spiral grooves are common on slip rings.

Multiple Brushes

If there are multiple brushes in a brush holder, each brush must carry its share of the amperes. Brushes will wear unevenly if they are not carrying equal current.

Open frame machines can be examined with an infrared gun. Take the temperature of each brush while the machine is running. A brush that is hotter than the rest may be carrying more than its share of the amperes.

A brush will arc and get hot if it doesn't have enough spring tension.

Brush Box and Commutator Spacing

A machine's brush boxes should be spaced at an equal distance from the commutator. The spacing should be close enough to keep the brush from chattering. This measurement can be from $1/16$ inch on small machines to $1/8$ inch on large machines. (Some large machines may specify even more space.)

A piece of plastic (the right thickness) inserted between the box and commutator works well for setting the box height. Plastic or a similar material won't scratch the commutator.

Distance Between Brushes

The distance between brushes around the commutator must be exactly the same.

Place a strip of paper (adding machine tape) under two adjacent sets of brushes. Pull the paper in the direction of the machine's rotation so that the brushes are against the trailing side of the brush box.

Mark the paper on the trailing side of each brush set. Compare this measurement to the rest of the brushes by inserting the tape under them the same way. A difference of as little as .05 inch can affect commutation.

Brush Holder Alignment

When there are multiple brushes per holder, all brushes should contact the same segments, unless designed otherwise. Misalignment will change the number of coils per pole, causing magnetic unbalance and brush arcing.

Spring Tension

All brushes must press the commutator with the same pressure. Brushes with more spring tension will carry more amperes. A scale designed to weigh spring pressure is available from brush companies (Fig. 2.4).

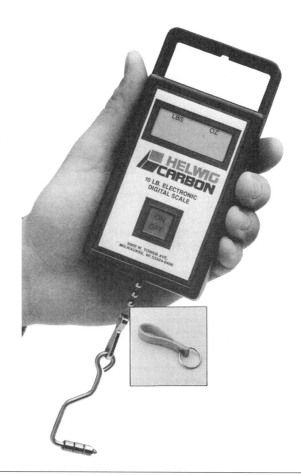

FIGURE 2.4   Scale used for exact measurement of brush spring tension.
Helwig Carbon Products, Inc.

The recommended spring pressure is as follows:

High-speed turbine rings and soft graphite brushes	2.25–2.75 lb
Normal speed slip rings	3.5–4.5 lb
Most industrial motors	4–5 lb
Metal graphite	4.5–5.5 lb
Fractional horsepower	5–8 lb
Traction motors	5–8 lb
Angled brushes with an angle greater than 25°	Add 1/2 lb

Brush weight must be considered when the machine has large brushes. The weight of the brush located on top is subtracted from this value, and the weight of the brush on the bottom is added to it. (The tension value of brushes located on the sides of the commutator is the same.)

Replacing adjustable spring brush boxes with constant spring pressure brush boxes has been very successful (Fig. 2.5).

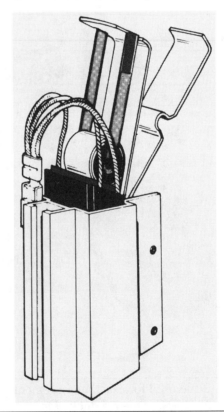

FIGURE 2.5   Continuous pressure brush holder. *Helwig Carbon Products, Inc.*

Brush Shunts

The brush shunt should be well anchored in the brush and show no discoloration from heat. The shunt wire must be large enough to carry the brushes' share of the amperes. (If the wire is discolored from heat, it isn't large enough.)

The shunt connection to the brush rigging must be clean and tight. Brushes should be changed before the shunt (or its components) rubs the commutator.

Side Wear

A brush with a poor shunt connection or a shunt wire that is too small will develop side wear on the trailing side of the brush.

The brush box and the shunt are connected to the same conductor. When the brush box conducts some of the amperes, side wear occurs.

The connection between the box and brush isn't solid; therefore normal brush movement will arc away brush material.

Uneven Commutator Segments

Uneven commutator segments will cause brushes to bounce and arc excessively. A faulty segment may lift as the armature turns but not be out of place when standing still. As a commutator heats to normal temperature, thermal growth can distort some segments. Chipped brushes are a symptom of a high segment.

Pressing on the top of the brush with a wood or fiber rod (while the machine is running) will detect uneven commutator segments.

Brush Dusting

Brush dust is a highly conductive contaminant that can cause short circuiting in the machine's circuitry.

Underloading (low brush current density) is the most common cause of brush dusting; another cause is low humidity. Brush and commutator wear is high without the lubrication that water vapor provides.

Low Load Current

Most brushes should carry 40 to 50 amperes per square inch of brush face area. If the load is lower than this, a different type of brush may be needed.

Generally speaking, gray brushes will handle underloading better than black brushes. Some can operate as low as 25 amperes per square inch. Peak amperes for the gray brush, however, would be 80 to 90 amperes per square inch—much lower than the black brush. Gray brushes are softer, hence less commutator wear. Black brushes can handle 110 to 125 peak amperes, but no lower than 40 to 50 amperes per square inch.

An aftermarket brush manufacturer should be consulted to solve brush problems if the machine's manufacturer can't provide information.

Use the following current density formula to determine normal current density. (First, substitute the motor's actual armature amperes for nameplate amperes to see if corrective measures should be taken.)

Current density = nameplate amperes ÷ ½ the brushes × width × thickness

Removing some brushes will lower the total brush area if the machine is underloaded, but the machine's complete duty cycle must be considered. (First, use the above formula to determine if this will work.)

Each brush holder must have the same number of brushes. The removed brushes must not be in line with each other. Figure 2.6 shows the right way to stagger the location of brushes.

Silicone Vapor and Adverse Ambient Conditions

Silicone vapor will cause brush dusting. Silicone grease should never be used on DC machine bearings for this reason. Forced ventilation (using outside air) is recommended when silicone contaminants can't be avoided. Sealing vent covers and replacing gaskets with silicone will cause silicone vapors, and should be avoided.

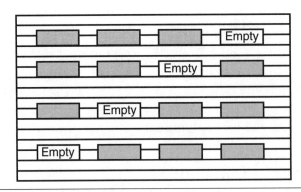

FIGURE 2.6 The correct location of removed brushes when increasing the current density (of remaining brushes).

Adverse ambient conditions affect the commutator and brushes. Air that contains caustic, corrosive, abrasive material or low humidity will shorten the life of the commutator and brushes. Clean air (with controlled humidity) will maximize the machine's running time.

Seating Brushes

Seating brushes to fit the commutator should be done carefully. The brush face must make good contact with all segments under it. Normal operation requires segments of one or more coils to be shorted together under a brush. If full contact between brush and segments isn't made, the magnetic neutral (explained in the next section) will shift, causing other brushes to arc. The motor will lose power and run hot. This will continue until the brushes are properly seated.

Thin sandpaper is used to do the initial seating (Fig. 2.7). It should be held against the commutator tightly, while rolling the armature or sliding the sandpaper back and forth. The final strokes should be in the direction of rotation. A seating stone can be used for the final seating.

Use a vacuum cleaner during and after this process to remove all residues. Covering the end of the commutator (risers) and the windings (up to the iron) with masking tape is recommended.

Brushes should move freely in the brush boxes and make good contact with the commutator. The brush face, which contacts the commutator, should be shiny. If any part of the brush face has a flat black color, it isn't making good contact with the commutator.

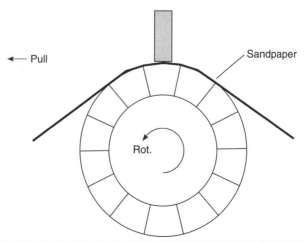

FIGURE 2.7   Seating a brush with sandpaper.

Neutral Location in the Stator and Armature

The neutral location in the stator is a spot exactly between its poles. It's 90 electrical degrees from each adjacent pole center. (Interpoles are placed in the neutral location.) The neutral spot in an armature iron is the teeth surrounding the slot or slots containing coils (that are shorted by a brush).

Whether using the DC kick method (see "Setting Neutral in a Generator") or applying AC to the shunt field, the neutral location is found by checking the commutator's bar-to-bar voltage. (The commutator will have little or no voltage bar to bar in the neutral location.) These bars are located where the voltage transformed into the armature windings from one stator pole is canceled by the voltage transformed by its adjacent stator pole.

When the machine is running, the brushes short the armature coil(s) located in this position. With idle coil(s) in this position, the armature teeth become a pole center, the same as with the pole iron of a shunt field coil (Fig. 2.8).

Setting Neutral in a Motor

Setting neutral in a motor can be done using a low-voltage voltmeter and 120 volts AC. (The AC voltage should be less than the shunt field's voltage rating.)

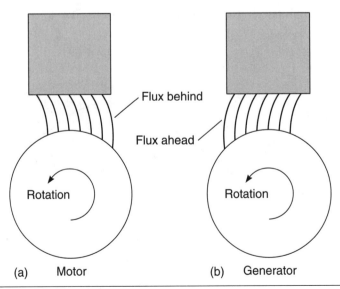

(a) Motor (b) Generator

FIGURE 2.8 Lines of force are bent under full load.

Attach the voltmeter to A1 and A2 (armature leads) and apply the AC voltage to F1 and F2 (shunt field leads). Move the brush holder back and forth until the lowest voltage reading is found. Secure the brush holder at this spot. Most machines will not have a 0 voltage spot. (Voltage under 0.5 volt is acceptable.)

Black Band Adjustment

If possible, check the brush arcing at full load. (Further adjusting may be necessary.) The flux (lines of force) will deform as shown in Fig. 2.8. Move the brush holder one segment at a time *against rotation* (Fig. 2.8a) until there is the least amount of sparking. This is called the black band method of adjusting the brushes. The term *black band* doesn't refer to a black band of any kind. Instead, it means the least amount of sparking, as opposed to the light that occurs with excessive sparking along the trailing edge of the brushes.

Setting Neutral in a Generator

To set neutral in a generator using the kick method, use a DC voltage source and a low-voltage voltmeter. The DC voltage should be no higher than the shunt field's rating. Attach the low-voltage voltmeter to A1 and A2. Attach the positive lead of the DC voltage source to F1. Make and break the connection between the negative lead and F2.

As the brush holder is moved, the lowest voltmeter reading will indicate the unloaded neutral spot. Secure the brush holder at this position.

Use the black band method (described in the previous section) for the final setting. If the brushes arc when the generator is loaded, move the brush holder *with rotation* one segment at a time until arcing is acceptable (Fig. 2.8b).

Restoring Residual Magnetism After Setting Neutral

AC voltage can be used to set the neutral in a generator, but the residual magnetism in the pole iron may become reversed, or demagnetized. If this method is used, apply DC voltage (no higher than the shunt field's rating) for a few minutes, with the positive lead on F1 and the negative lead on F2. This will restore the residual magnetism in the pole iron to the right polarity.

A generator can be tested by running it as a motor. This test method will invariably reverse the residual magnetism in the generator's pole iron. Use the preceding procedure to restore the right polarity in the pole iron.

Overheated Motor

When a motor has overheated to the point of smoking, it's best to replace or rewind it. It should be inspected very carefully for burned windings. There will be a very distinct burned winding smell (that stays in the motor after it cools). The components that overheat are all in the armature circuit—the armature, the interpoles, and the series field. The shunt field may be damaged if the series field is wound on top of it.

Sometimes an overheated motor will operate for a long time. Therefore, it should be tested more frequently than normal. The fact that it was overheated should be recorded.

Field Testing the DC Machine

The term *machine* is used when both motors and generators use the same procedure. *Motor* or *generator* is used when the explanation applies to one or the other.

The following test procedures are designed to quickly test DC machines in place. These tests identify the problem and determine if the machine should be removed, can be repaired in place, or can continue to operate as is. The text describes the circuitry of a compound machine. The shunt field and armature use the same test, respectively, in all types of machines. A series motor's field coils may have many more turns than the series field of a compound machine and can be tested like a shunt winding.

All test information should be recorded.

Field Testing the Armature

The right size ohmmeter or a limited DC current and an ammeter can be used to check the armature (while the machine is in place). This test can be done from the machine's control.

Connect the test equipment to A1 and A2. Rotate the armature slowly, noting the reading. The value will change as the armature rotates. There will be a high and low reading. It may vary (but should repeat the same variance) during a complete revolution. (Any change indicates a problem and requires more detailed testing.)

An armature can be tested in place to determine if a motor needs to be removed and repaired. Uncouple the load and lift all the brushes. Apply full-rated voltage to the shunt field, and rotate the shaft. There will be more drag when the shunt field is energized. If the drag is hard and uneven, there is a short in the armature winding.

Shorted turns form a closed-loop circuit. The turns within this circuit generate a voltage and current as they cut the shunt field lines of force. The power that the loop circuit creates requires torque, the same as a generator. The retarding torque is uneven because it happens only as the shorted turns pass a stator pole.

This test can be simulated using a motor that has no problem. Lift the brushes and short two commutator bars together. Turn the shaft before and after applying rated voltage to the shunt field. This procedure gives a feel for what to expect when field testing a real short.

Field Testing for Grounds in the Armature Circuit with an Ohmmeter

Leads A1 and A2 include the brush holder and interpoles. A test for ground should be done from A1 or A2 to the shaft and to the frame with a multimeter set on ohms. The machine's bearings shouldn't be included in a test circuit using amperes (such as a microhmmeter has).

A solid ground in the armature would require rewinding. An indication of 1 megohm or less requires further testing.

A low reading of 1 to 50 megohms can sometimes be increased by cleaning and drying the machine.

Field Testing the Shunt Field

All test information should be recorded and available for future comparison. The shunt field leads F1 and F2 should be disconnected from the armature circuit for the following tests. Problems in the shunt field are

- Ground to frame
- Shorted turns
- Open shunt field
- Shorted to the series field

Ground to the Frame Symptoms

The following are ground to the frame symptoms:

- Loss of power
- Higher RPM with no load
- Some or all brushes arcing
- Frequent ball bearing failure
- Higher than nameplate amperes if more than one coil is shorted to ground
- Some shunt field coils get hotter than others

Field Test Using an Ohmmeter to Test for Ground in Shunt Field   Be sure the power is disconnected and that shunt field leads F1 and F2 aren't connected to the armature circuit.

Secure one ohmmeter probe to the machine's frame, making good electrical contact. Then, touch F1 with the other probe. No problem would read near infinity.

If the reading is from 1 to 50 megohms, the problem could be moisture. The field coils should be cleaned, dried, and retested. Extreme amounts of moisture affect the coils located at the bottom of the machine first.

If the coils are loose on the pole iron, move each coil to locate the ground. Field coils should be tightly secured on the pole iron. Loose coils eventually wear through the insulation, causing electrical contact to the pole iron.

Shorted Turns Symptoms

The following are shorted turns symptoms:

- Higher than nameplate amperes
- Shunt field coils hot
- Higher than normal no load speed
- Some or all brushes arcing excessively
- Lower starting torque

Field Test for a Shorted Shunt Field Using an Ohmmeter   Be sure the power is disconnected and that shunt field leads F1 and F2 are not connected to the armature circuit. Record the ohmmeter reading across F1 and F2.

Compare this reading to the shunt field resistance given on the nameplate. If the resistance isn't given, divide the shunt field voltage by its amperes. If the field coils aren't shorted, the result should be very close to the ohmmeter value recorded across F1 and F2.

Use the following procedure if the resistance of each coil is required: Measure the applied test voltage, and divide it by the resulting amperes (to get total resistance). Divide the resistance by the number of coils in the circuit to get the resistance per coil.

If there's more than one circuit, multiply the total resistance by the number of circuits. The result will be the resistance per circuit. Divide the resistance per circuit by its number of coils to get the resistance per coil.

The resistance of the coils is lower when they are cold than when at running temperature. Nameplate values are determined at running temperature.

A reading 5 percent or more lower than the expected value indicates the shunt field coils should be tested further.

Field Test Using an Ammeter to Test a Shorted Shunt Field   Apply the nameplate-rated DC voltage to the shunt field leads F1 and F2. There could be shorted turns if the reading is above the shunt field's (nameplate) amp rating.

Allow the field to warm to its normal running temperature if the amperes are 5 percent or less above the nameplate value. (Nameplate amperes assume the machine is at normal running temperature.) Copper gains resistance as its heat increases. The amperes will drop to normal as the machine reaches its operating temperature.

Open Shunt Field Symptoms

The following are open shunt field symptoms:

- Very little starting torque
- Overspeeding
- Brushes arc excessively
- High amperes in armature circuit at any speed

 If the shunt field circuit is open while the motor is running with no load, it will dangerously overspeed and damage or destroy the armature components.

Field Test with an Ohmmeter for an Open Circuit in a One-Circuit Shunt Field   Make sure the power is disconnected and that the shunt field leads F1 and F2 aren't connected to the armature circuit. There will be no ohmmeter reading (infinity) between F1 and F2 if the circuit is open.

Test to the frame with both F1 and F2 for a ground.

Field Test with an Ohmmeter for an Open Circuit in a Parallel Circuit Shunt Field   Make sure the power is disconnected and that the shunt field leads F1 and F2 aren't connected to the armature circuit.

If the machine has more than one circuit in its shunt field, an open would have more resistance than normal between F1 and F2 (Fig. 2.9).

Determine the resistance of the shunt field. (If the resistance is not given on the nameplate, divide the shunt field voltage by its amperes.) If tests from F1 to F2 show a higher resistance than expected, one of the circuits is open. A two-circuit connection would have double the expected resistance. If there were three circuits, the resistance would be 1.5 times the expected resistance. (The more circuits in the shunt field, the smaller the difference.)

Test to the frame with both F1 and F2 for a ground.

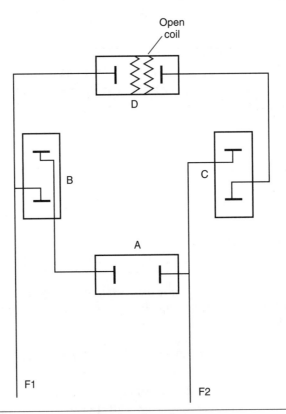

FIGURE 2.9   Two-circuit shunt field with coil D open.

Shunt Field Shorted to the Series Field Symptoms

The following are shunt field shorted to the series field symptoms:

- High shunt field amperes
- Some hot coils
- Erratic speed

Field Test with an Ohmmeter for the Shunt Field Shorted to the Series Field   Make sure the power is disconnected and that the shunt field leads F1 and F2 are not connected to any other circuit. If there is no short, a test from F1 to S1 will read infinity.

In smaller machines, the series coil is wound on top of the shunt coil, with insulation separating them. Overloading is a common cause of insulation breakdown between the two fields (Fig. 2.10). An overload overheats the series field, deteriorating and eventually charring the separating insulation. This problem requires rewinding all the field coils.

If the machine hasn't been overheated and the reading is 50 megohms or less, cleaning and drying should make an improvement.

Field Testing the Series Field

For this testing, the series field leads (S1 and S2) are connected in series with the armature and interpoles.

Problems in the series field include

- Ground to frame
- Shorted turns
- Open
- Series field shorted to the shunt field (also covered under shunt field problems)

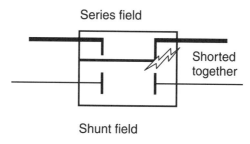

FIGURE 2.10   Shunt field shorted to the series field.

Ground to the Frame Symptoms

The following are the symptoms of a ground to the frame:

- No symptoms (unless more than one coil is grounded)
- Touching the machine gives a shock
- High amperes in armature
- Loss of power
- Arcing at some brushes
- Frequent ball bearing failure

Field Test Using an Ohmmeter to Test for Ground in Series Field   The series field coil is usually wound on top of the shunt field coil. In some machines, it's a separate coil.

Be sure the power is disconnected. Disconnect the series field leads S1 and S2.

Secure one ohmmeter probe to the machine's frame, making good electrical contact. Then, touch S1 with the other probe. No problem would read near infinity. A reading (nearly the same as touching the probes together) indicates a serious ground.

If the coils are loose on the pole iron, a faulty coil may not make continuous contact to the frame. Move all the coils while the test probes are in place. (Field coils should be tightly secured on the pole iron and have no movement.) Loose coils eventually wear through the insulation, causing electrical contact to the pole iron. If there is a low reading of 50 megohms or less, cleaning and drying should make an improvement.

Shorted Turns in Series Field Symptoms

The following symptoms indicate shorted turns in the series field:

- Loss of power output
- Arcing at some or all brushes
- Output not responding to load changes

Field Test for Shorted Turns in Series Field   The series field is a high-ampere circuit (with large wire). The location of an open, a ground, or shorted turns is usually very obvious. When a fault occurs, there is major arcing, melted wire, and scorched insulation in the fault area, as seen in Fig. 2.11.

Melted compensating
winding bars

FIGURE 2.11   This short in the series field also melted the compensating winding. *P&H MinePro Services.*

A shunt field with the same faults will not cause as much damage. Its small wire will melt and open its circuit, with fewer amperes and less arcing.

Operate the machine loaded (with the series field) and check for symptoms described above. Next, operate the machine without the series field. It will now operate as a shunt-wound machine.

If the machine is a motor, the RPM will drop some when the load is applied. If it's a generator, the voltage will drop some as the load is applied. Brush arcing will be normal.

If the machine's series field is multicircuit, there will be a large magnetic imbalance. Some (not all) brushes will arc excessively. Arcing will increase as the load increases.

Shorted turns in the series field are hard to locate without opening the machine. The voltage drop test across each coil is the surest way to detect this problem. This requires accessing the connections between coils.

Open Series Field Symptoms

The following symptoms indicate an open series field:

- One-circuit series field
  - Motor: Will not run.
  - Generator: No voltage across A1 and S1. There will be full voltage across A1 and A2.

- Multicircuit series field
  - Motor: Runs normally with no load, but RPM drops and some brushes arc excessively when loaded.
  - Generator: Full-voltage output with no load but voltage drops and some brushes arc excessively when loaded.

Field Test for Open Series Field by Operating Without the Series Field   Disconnect S1 and S2. Operate the machine as a shunt motor or generator. It should operate as a shunt machine, without the stabilizing effect of the series field.

A motor will slow down as the load is added. A generator's voltage output will drop as the load is added.

Field Test with an Ohmmeter for Open Series Field   Be sure the power is disconnected and that series field leads S1 and S2 aren't connected to any other circuit. A one-circuit series field will read infinity across S1 and S2.

A multicircuit series field requires disassembling the machine and opening the connections. This is described later in the chapter under "Locating an Open Circuit in Series Field."

Series Field Shorted to the Shunt Field

A test between F1 and S1 with an ohmmeter should read infinity. A reading as high as 1 megohm indicates a potentially serious problem.

The series field is wound on top of the shunt field with a layer of insulation between. This insulation must withstand the pressure of winding large series coil wire tightly around the shunt coil. It also must withstand the thermal growth stress caused by heating and cooling of the series coil. An insulation breakdown usually blows the shunt field open, and makes a very visible burned spot on the coil.

The voltage difference between the two circuits can be the value of the applied voltage.

Locating Problems in the DC Machine

The following tests locate the problem and recommend the extent of repair to be done. Most of the procedures require dismantling the machine to access the components.

All test information should be recorded for future reference.

Locating Problems in the Armature

Ohmmeter Test for Ground in the Armature Circuit

The leads A1 and A2 include the brush holder and the interpoles. Test for a ground from A1 to the frame with an ohmmeter. If the reading is low (from 1 to 50 megohms), cleaning and drying may be all that is needed.

It's normal for the armature to have a low megohm reading to the shaft (1 to 50 megohms). Brush dust, which is highly conductive, accumulates behind the commutator during normal operation. This is why DC power is not grounded.

If a ground develops in a machine (for example, a locomotive), all motors on the machine have a common connection to each other. The first indication of a ground is bearing failure. (Leakage current from the commutator travels through the bearings.)

Use the following procedure to locate the problem: Lift the brushes and check from the commutator to the shaft with the ohmmeter. A solid ground in the armature (as seen in Fig. 2.12) would require rewinding. If no problem is found in the armature, test the brush boxes—both polarities or adjacent brush boxes—to the frame (Fig. 2.13).

If the brush box connected to the interpoles has a low reading, disconnect the interpoles and test again. If a low reading is in the interpoles, cleaning

Melted wire
and iron

FIGURE 2.12   Armature with its winding melted and fused to the iron.
P&H MinePro Services.

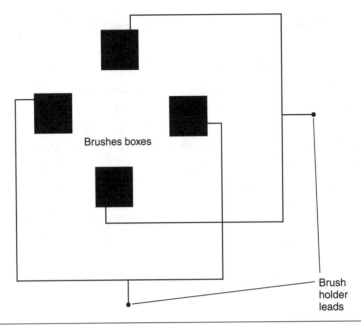

Brushes boxes

Brush holder leads

FIGURE 2.13   Every other brush holder is the same polarity; adjacent brush holders are the opposite polarity. The brush holder (with the brushes removed) should be isolated from all circuits and from the machine's frame.

and drying them should improve the reading. A solid ground requires replacing or reinsulating the coil(s).

If the brush holder has a low reading, clean and/or replace its bushings and insulators.

Heat-Damaged Interpoles

Interpoles, like the series field, are in series with the armature. If the armature has been overloaded or shorted, high amperes have very likely overheated the interpole coils. If the coils are discolored, they should be replaced or reinsulated.

Locating a Grounded Interpole with AC and a Clamp-on Ammeter

For safety reasons, a low-voltage transformer (with high-ampere capacity) and a current-limiting resistor should be used. This will isolate the DC machine from the equipment ground circuit of the AC power source. The transformer should have enough power to provide a readable number of amperes. A solder gun is sometimes used but can overheat.

Fasten one line of the voltage source to the frame and the other to an interpole lead. Apply a small number of amperes to the circuit. With a clamp-on ammeter, check the amperes at each coil-to-coil connection. The connection that shows no reading (Fig. 2.14) will be beyond the faulty coil. Disconnect this coil, and check for a ground between it and the previously tested coils. If the end coil contains the ground, feed the circuit from the brush holder connection, and proceed toward A1.

The decision to repair or rewind a single interpole will depend on the age and condition of the machine.

If any series circuit components show signs of heat damage from high current, all coils in the circuit should be rewound or reinsulated.

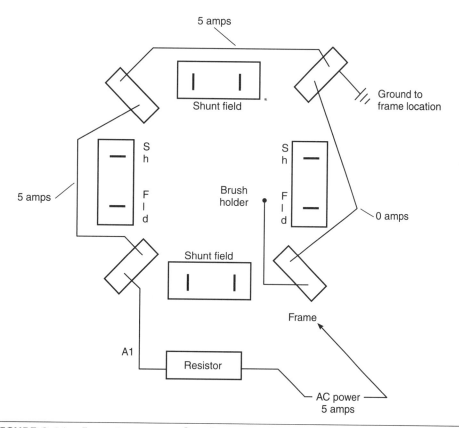

FIGURE 2.14   Locating a ground in the interpole circuit with a clamp-on AC ammeter and limited AC current.

Locating a Shorted Interpole Using the AC Voltage Drop Test

A 120-volt AC voltage source with a current-limiting resistor and a voltmeter can be used for this voltage drop test. Limit the amperes to the armature nameplate rating or less. AC works better because of inductance, which is explained in Chapter 3.

 The armature should not be included in the AC test circuit.

At least two interpole coils in series are needed for this test method. The coils are compared, as illustrated in Fig. 2.15. Take a voltage drop test across each coil. (A lower voltage is found across a shorted coil.)

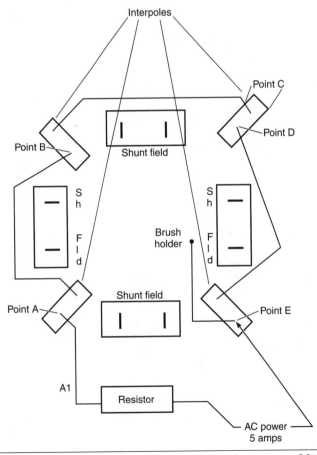

FIGURE 2.15   Locating a shorted interpole with a voltmeter and limited AC current.

This voltage drop test could result in the following "smoke test" if enough amperes are used.

Locating a Shorted Interpole Using the Smoke Test

A shorted turn or two can be very hard to detect when large wire is used in the interpole coils.

The smoke test can be used to locate shorted turns. Use 120 volts AC for the test. Limit the amperes to the armature circuit's nameplate value with a resistor. (Make sure the armature isn't included in the test circuit.)

A high circulating current will be transformed into the turns within a short, causing them to smoke almost instantly. Testing should be stopped before the rest of the coils overheat.

If the coils are constructed of flat wire, they can be reinsulated (if they aren't melted or fused together).

Locating Problems in the Shunt Field

The following troubleshooting strategy is applicable to both the shunt machine and the shunt winding of a compound machine.

The Comparison Test

The best method for testing the shunt field (F1 and F2) is to compare one coil's test data to another coil's test data. There's little chance that all the coils would have the same number of turns shorted.

Exception: If all the coils give the same test results and the amperes are too high or too low, the winding data may have been redesigned to operate on a different voltage. This is sometimes done when the shunt field is separately excited.

Needle Probes

If the test instruments are a standard ohmmeter or a voltmeter, a sharp needle probe may be used to puncture the insulation covering a coil-to-coil connection (Fig. 2.16). Neither instrument will cause a high enough ampere flow to damage the small probes. If no problem is found, the puncture can be varnished or wrapped with tape.

FIGURE 2.16 Needle probes for piercing insulation using a voltmeter or an ohmmeter. Coils and circuits can be tested quickly without removing the insulation. *Fluke.*

Variations in the Shunt Field Circuitry

It's important to understand the shunt field's internal connection before testing it:

- All coils being tested must be alike.
- There may be more than one circuit in the shunt field.
- Some machines have more than one shunt field coil per pole.

Drawing Simple Schematics of a Shunt Field

Figure 2.17a is a DC shunt machine and Fig. 2.17b is a simplified drawing of this machine's connection. (It's a one-circuit shunt field.) The arrows in Fig. 2.17b show where the test probes should contact the connections.

Drawing a Diagram of a Two-Circuit Shunt Field

Figure 2.18a is an example of a two-circuit shunt machine. Figure 2.18b is a simplified drawing of this machine's connection. (It is a two-circuit shunt field.) Each circuit has two coils in series.

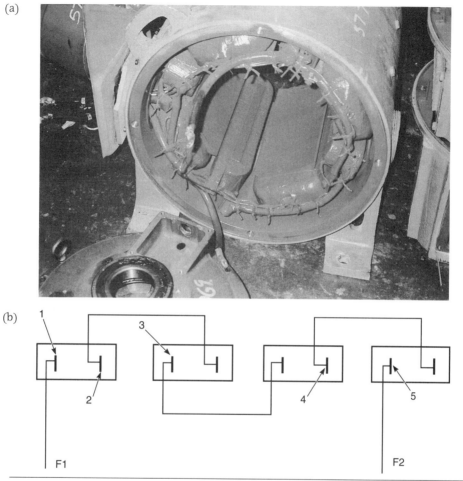

FIGURE 2.17   (a) The connection side of a one-circuit shunt field stator.
(b) Sketch of the shunt field connections in (a).

Drawing a Diagram of a Two-Circuit, Two-Coils-per-Pole Shunt Field

Figure 2.19a is an example of a two-circuit, two-coils-per-pole shunt machine. Figure 2.19b is a simplified drawing of this machine's connection. Each circuit has four coils in series. Problems in the shunt field are

- Ground to the frame
- Shorted turns

(a)

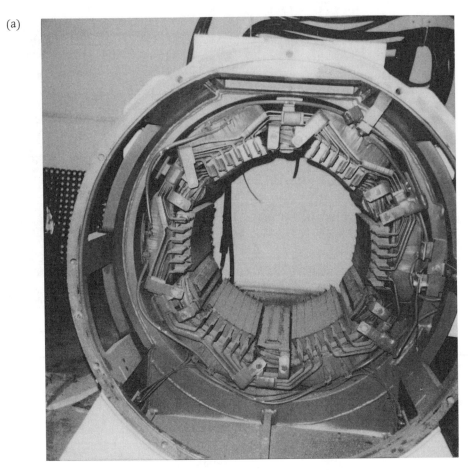

(b)

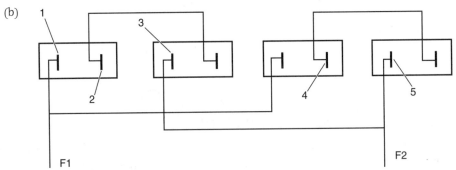

FIGURE 2.18   (a) The connection side of a two-circuit shunt field stator. (b) Sketch of the shunt field connections in (a).

(a)

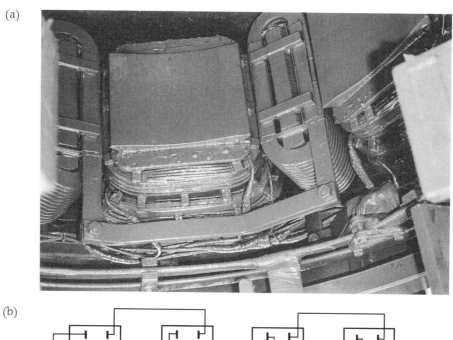

(b)

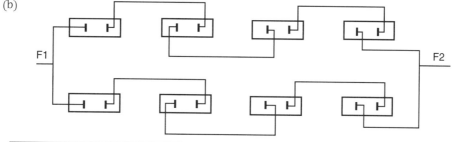

FIGURE 2.19   (a) The connection side of a two-circuit shunt field stator. *P&H MinePro Services.* (b) Sketch of the shunt field connections in (a).

- Open circuit and shunt field
- Shorted to the series field

Ground to the Frame Symptom

Frequent ball bearing failure.

Ground to Frame Symptoms When More Than One Coil Is Shorted to the Frame

- Higher than nameplate amperes in the shunt field
- Some shunt field coils get hotter than others

- Arcing at some brushes
- A motor will lose power and overspeed (with no load)
- A generator won't develop full voltage and power output

Locating a Grounded Shunt Field Coil Using an Ohmmeter

Separate F1 and F2 from the other leads. Check the resistance of the shunt field with an ohmmeter. A value 25 percent less than the shunt field-rated resistance indicates more than one coil is grounded—or there are shorted turns.

Secure one probe of the ohmmeter to the frame of the motor, making good electrical contact. Touch F1 with the other probe.

No problem will read near infinity. A reading less than the shunt field's total resistance indicates a grounded coil. A very low reading means the ground is close to the lead being touched. A reading close to the resistance of the shunt field indicates the ground is near the other lead.

Remove the insulation from the coil-to-coil connections, and test each coil to the frame. The coil with the least resistance to the frame will contain the ground. Disconnect this coil and verify that it contains the ground. Test the rest of the shunt field coils for a ground.

When two coils short to the frame, at least one coil will be bypassed. The result is high amperes, and all but the bypassed coils will get hot.

The coils should be tight on the pole iron. If not, move them with the ohmmeter attached to the frame and F1.

Shunt field coils are wrapped with insulation and are very well insulated from the frame. Brush dust cannot penetrate this insulation. If the lead-to-frame reading is less than 50 megohms, the coils are probably damp and should be cleaned, heated, and dried out.

If there is dampness, check the coil in the lowest location. Disconnect the bottom coil to determine if that is causing the low reading. If cleaning and drying do not produce a near-infinity reading, replace the coil. (A single grounded coil is one of the few cases in which replacing only one coil is recommended.)

DC machines should not have an equipment ground. The bare copper bars of the commutator (combined with highly conductive brush dust) make a high-leakage circuit. If a shunt field coil has a ground, the leakage current will go through the bearings to the commutator, resulting in frequent bearing failure and commutator damage.

Shorted Turns in the Shunt Field Symptoms

The following symptoms indicate shorted turns in the shunt field:

- High amperes
- Shunt field coils hot
- Some or all brushes arc excessively
- A motor will have low starting torque and higher than normal no-load speed
- A generator will not develop full voltage and power when loaded

Locating a Shorted Shunt Field Coil Using an Ohmmeter

Be sure the power is disconnected and that one of the shunt field leads (F1 or F2) isn't connected to the armature circuit. Record the ohmmeter reading across each coil. (There should be a difference of no more than ±2 percent.)

Locating a Shorted Shunt Field Coil Using a Voltmeter

Apply nameplate-rated DC voltage to the shunt field (F1 and F2). If necessary, use a resistor to limit the amperes (to no more than 10 percent higher than shunt field nameplate amperes).

Divide the nameplate shunt field voltage by the number of coils in series to find the normal voltage across each coil. If all the coils are good, the voltage across each coil will be within 5 percent of each other.

A shorted coil will have lower voltage across it than the rest of the coils. When the shunt field's rated voltage is applied, the difference between a shorted coil voltage and normal coil voltage is split among the rest of the coils. The voltage across each coil is increased (which raises the amperes and increases the heat).

Verifying That a Shunt Field Coil Is Shorted Using the Smoke Test

A coil that has a 4 or 5 percent lower ohmmeter reading than the rest may or may not have shorted turns. The smoke test will determine whether there are any. (This test won't damage a coil that isn't shorted.)

Open the coil's connections and apply AC voltage to them. The voltage should be limited to the shunt field's nameplate voltage. (Turns within the short will get hot, indicating the coil needs to be replaced.)

If the machine has run (with high amperes and temperature) long enough to damage the coil insulation, all coils in the circuit should be replaced.

Locating a Shorted Coil in a Parallel Circuit Shunt Field with an Ohmmeter

Figure 2.20 shows this motor's connection. Two shunt field circuits in parallel should have identical amperes. A circuit with higher amperes than the others will have a shorted coil(s).

When testing from a to b in Fig. 2.20, the ohmmeter will show less resistance than this coil has. (This is because of the loop circuit through the remaining coils in the parallel circuit.) Although the loop circuit's resistance isn't the resistance of the coil, this value can be compared with all coils in the two circuits. If there are no shorted coils, their values will be within 2 percent of each other.

Locating a Shorted Coil in a Parallel Circuit Using a Voltmeter

If a voltage drop test is used on the machine in Fig. 2.20, each coil should read within 5 percent of the others. Voltage across each coil in this example should be approximately one-fourth the applied voltage. If coil x has shorted turns, the voltage across the rest of the coils in this circuit will be higher than one-fourth the applied voltage. (All coils in this circuit should be replaced.)

The following are open shunt field symptoms:

- Very little starting torque
- Overspeeding
- Brushes arc excessively
- High amperes in armature circuit at any speed

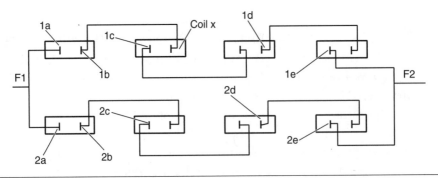

FIGURE 2.20   A drawing of a two-circuit shunt field, with the points for testing indicated.

Shunt Field with an Open Circuit

The motor won't start a load if the shunt field circuit is open. If it's running when the open occurs, the control should shut it off immediately.

A loaded motor will try to overspeed. It will lose most of its counter-voltage, and the amperes through the armature will become excessively high. If there is no load, the motor will accelerate until the armature is destroyed.

When test running an unloaded motor with a shunt field, be sure F1 and F2 are securely connected.

Locating an Open Circuit Using an Ohmmeter

The motor in this test has a one-circuit shunt field (Fig. 2.21). There is no ohmmeter reading between F1 and F2. Connect one ohmmeter lead securely to F1. Pierce the insulation as shown in Fig. 2.21 at point #5. (This is on the coil side of the connection from lead F2.) If there is no reading, proceed to point 4, etc., until a reading is found.

If the open coil was shorted before it opened, all coils in the circuit have probably overheated. In this case, they should all be replaced.

The only exceptions to replacing all the coils are an open coil that has worn through the insulation (from being loose on the iron) or a coil that has been physically damaged.

Locating an Open in a Parallel Circuit Using an Ohmmeter

Figure 2.22 shows a two-circuit shunt field. With the nameplate information for the shunt field being 100 volts and 10 amperes, use the following procedure to determine the resistance of a shunt field coil:

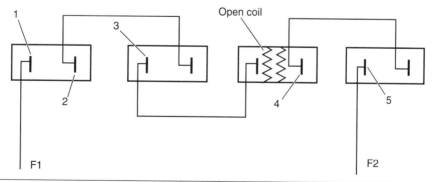

FIGURE 2.21 A one-circuit shunt field with an open coil.

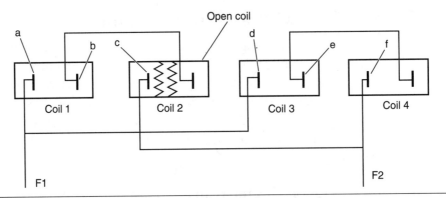

FIGURE 2.22   A two-circuit shunt field with an open coil.

100 volts ÷ 10 amperes = 10 ohms (the resistance of the shunt field)
10 ohms × 2 circuits = 20 ohms per circuit

   There are two coils per circuit:

20 ohms ÷ 2 coils = 10 ohms per coil

   A test from F1 to F2 would show 20 ohms. This is twice the shunt field's normal resistance.

   First, the connections should be checked carefully. A faulty connection will get hot, become charred, and eventually open completely.

   In Fig. 2.23 points a, b, c, d, e, and f are internal connections between coils and leads.

   If coil 2 is open, a test from a to b will show the resistance of coil 1 only (10 ohms). The same value would be found from d to e and from e to f.

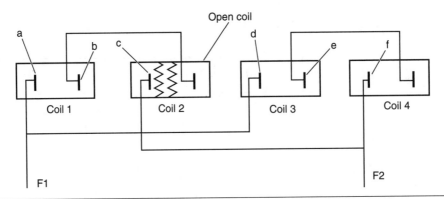

FIGURE 2.23   Test points for locating an open circuit in a two-circuit shunt field, using an ohmmeter.

Across open coil 2, b to c, the circuit would be through coils 1, 3, and 4. The result would be the resistance of three coils added together, or 30 ohms.

A high resistance reading would identify the open coil in a parallel connected shunt field.

Locating an Open in a Parallel Circuit Using a Voltmeter

The shunt field in Fig. 2.24 is rated 100 volts and 10 amps. Coil 2 is open. If 100 volts are applied to F1 and F2, it will draw 5 amperes, one-half the nameplate rating. Piercing the connections at point a and b with the voltmeter probes will show no voltage. Points d to e and e to f will read 50 volts, or one-half the applied voltage. Point b to point c will read full voltage, identifying the open coil.

Shunt Field Shorted to the Series Field Symptoms

The following symptoms indicate a shunt field shorted to the series field:

- High shunt field amperes
- Some hot shunt field coils
- Erratic speed

Locating Shunt Field Shorted to the Series Field Using an Ohmmeter

Normally there are separate leads for accessing the armature and series field. When a machine is custom made for a definite load, the two circuits may have an internal connection that eliminates one of the shunt field leads.

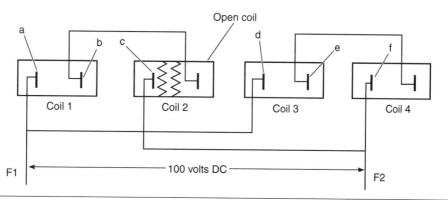

FIGURE 2.24 Locating an open circuit in a two-circuit shunt field with a voltmeter, and the field's rated voltage.

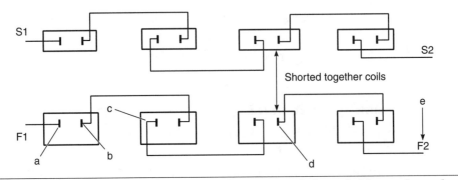

FIGURE 2.25 Locating the short when the series and shunt fields are shorted together (using an ohmmeter).

Be sure the power is disconnected and disconnect the field leads. If there is no short, a test from F1 to S1 will read infinity.

Figure 2.25 illustrates a coil-to-coil short in coil 3. Connect one ohmmeter probe to S1. With the other probe, pierce the shunt field connections at points a, b, c, d, and e. Points c and d will have the lowest readings, indicating the short is located in the coil between them.

A shunt and series field insulation breakdown is usually caused by burned insulation between the fields. The load amperes flow through the series field. An overload overheats the series field, deteriorating (and eventually destroying) the separating insulation. This problem requires rewinding all field coils.

Locating Problems in the Series Field

If the brushes arc excessively and a faulty coil in the series field is suspected, run the machine without the series field. It will then operate as a shunt machine. (In a motor, RPM will drop with load increase; with a generator, voltage output will drop with load increase.) If brush arcing is normal, the series field has a problem.

The series field (S1 and S2) is a low-resistance circuit, with very few turns. Anywhere from 5 to 50 turns of large wire are common. The counter-voltage in the armature controls the amperes through this circuit in a motor.

The following problems occur in the series field:

- Ground to frame
- Shorted turns

- Open circuit
- Short to the shunt field (covered earlier in the chapter under "Field Testing the Shunt Field")

Ground to Frame Symptoms

- No symptoms
- Electrical shock
- Frequent ball bearing failure

If more than one coil is shorted to ground, the symptoms will be arcing at some or all brushes, and some series field coils will get hotter than others.

A motor will have:

- Loss of power
- Lower RPM (loaded)
- Armature amperes higher than nameplate
- Premature bearing failure

In a generator, the following will occur:

- Low voltage/power output
- Premature bearing failure

Locating a Ground in the Series Field with a Clamp-On Ammeter and Limited AC

The AC voltage must be limited to 5 volts or under. A transformer and a (current-limiting) resistor should be used. The transformer should have enough power to provide a readable number of amperes.

Connect one transformer lead to the frame and the other to S1 (Fig. 2.26). Put the ammeter clamp around the coil-to-coil connections, starting at the nearest coil to S2. Check each connection until amperes show on the meter. Disconnect the faulty coil and check the remaining coils.

Locating a Ground in the Series Field with an Ohmmeter

Disconnect S1 and S2. Secure one probe of the ohmmeter to the frame of the machine, making good electrical contact. Fasten S1 to the other probe. If there is a low reading, open the coil-to-coil connections, starting at the

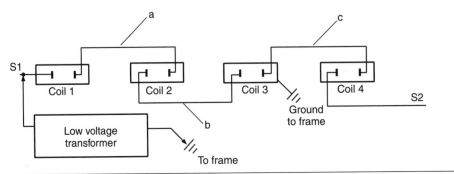

FIGURE 2.26   Locating a ground in the series field with low-voltage AC and a clamp-on ammeter.

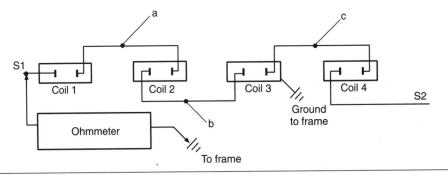

FIGURE 2.27   Locating a ground by opening coil-to-coil connections.

nearest coil to S2 (Fig. 2.27). When the reading goes to infinity, the ground is located in the last coil disconnected.

When two coils short to the frame, the turns (or coils) between them are bypassed. Bypassed turns or coils cause a severely unbalanced magnetic field. This causes some or all of the brushes to arc when the machine is loaded. If the coils aren't tight on the pole iron, move all of them, with the ohmmeter attached to the frame and a lead. If the lead-to-frame reading is less than 50 megohms, the fields should be cleaned and/or heated and dried out.

Shorted Turns in the Series Field Symptoms

If the turns are shorted in the series field, the following can occur:

- Some or all brushes arc when the machine is loaded
- Loss of power
- Machine doesn't react as it should to load changes

Locating Shorted Turns in the Series Field Using AC and the Shunt Field

Apply 110 volts AC to F1 and F2 (Fig. 2.28). Pierce series field coil-to-coil connections, and read the voltage across each coil. A lower reading will be found across a shorted coil. A circulating current in the shorted turns of the series coil may raise the amperes in the shunt field. Don't allow the shunt field to overheat.

Locating Shorted Turns Using the Voltage Drop Test and AC

An ohmmeter will not detect any difference between coils when only a few turns are shorted out (because of the circuit's low resistance). A microhmmeter will work; the voltage drop test (using 2 to 5 volts AC) also will work (Fig. 2.29).

 Never apply more than 2 to 5 volts AC to the series field (S1 and S2). A destructively high voltage will be transformed into the shunt field.

Remove the insulation from the series field coil-to-coil connections. Apply 2 to 5 volts AC to S1 and S2. Read the voltage across each series field coil. A lower reading will be found across the shorted coil.

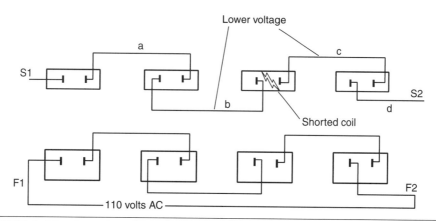

FIGURE 2.28   Locating shorted turns in the series field using the shunt field as a transformer.

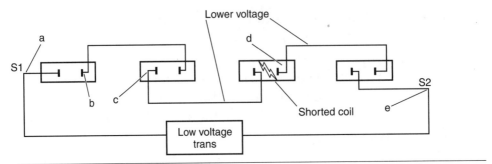

FIGURE 2.29 Locating a shorted coil in the series field using low-voltage AC and the voltage drop test.

Locating an Open Circuit in Series Field

A one-circuit series field with an open circuit will have the following symptoms:

- A motor won't start. The only complete circuit will be the shunt field.
- A generator won't have a voltage output. A1 and A2 will read full voltage.

A two-circuit series field with an open circuit will have these symptoms:

- A motor will lose power, and some or all of the brushes will arc excessively (from magnetic unbalance).
- A generator will have full voltage output (at no load). As it is loaded, the voltage output will drop some, and some or all brushes will arc excessively.

Locating an Open in a One-Circuit Series Field Using an Ohmmeter

An ohmmeter connected to S1 and S2 will read infinity (Fig. 2.30). Starting with an ohmmeter probe connected to S1, pierce each coil connection from S1 to S2 until there is no reading. The reading across the faulty coil will be infinity.

Locating an Open in a Two-Circuit Series Field Using a Voltmeter

Remove the insulation from the series field coil-to-coil connections. Use a resistor that limits the amperes to the armature's nameplate value, and apply DC voltage to S1 and S2 (Fig. 2.31).

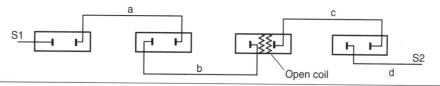

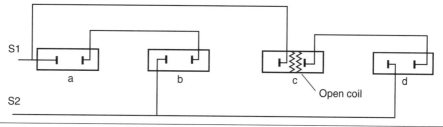

FIGURE 2.30   Locating an open coil in a single-circuit shunt field with an ohmmeter.

FIGURE 2.31   Locating an open coil in a two-circuit series field using limited current DC and a volt-meter.

 Never apply the motor's rated voltage to S1 and S2.

Check across each coil with a voltmeter. Full voltage will be found across the open coil. Coils a and b will show half voltage. Coil c will read full voltage and coil d will read zero voltage.

Identifying Leads of a Compound DC Motor with an Ohmmeter

Most compound DC motors have three circuits, the armature (A1 and A2), the series field (S1 and S2), and the shunt field (F1 and F2).

Use an ohmmeter to locate the three pairs (that have a circuit to each other). The shunt field leads (F1 and F2) are usually smaller than the armature and series field leads. The shunt field has much higher resistance than the armature and series field.

Identify and temporarily label the shunt field leads F1 and F2. Identify the armature leads by touching one ohmmeter probe to a lead and the other to the commutator. Label them A1 and A2. Temporarily label the remaining leads S1 and S2.

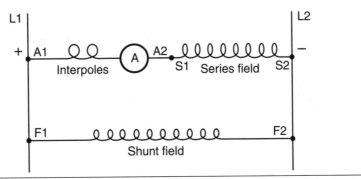

FIGURE 2.32 Identifying unmarked leads to conform to NEMA standards.

Connect the motor as shown in Fig. 2.32 and apply enough DC power to make it rotate. The rotation should be counterclockwise facing the commutator. Interchange A1 and A2 if the rotation is wrong.

Disconnect F1 and apply enough power to A1 and S2 to get rotation. If rotation reverses, interchange and re-label the series field leads.

NEMA standard rotation is counterclockwise facing the commutator end, with positive on A1 and F1.

Chapter 2 Review

1. Most of a DC machine's problems occur in the armature (page 57).
   T____ F____

2. Lacquer thinner is a good liquid for cleaning a commutator (page 58).
   T____ F____

3. An open armature coil may cause dead spots (page 58). T____ F____

4. Burned commutator segments spaced equal to the distance of two like-polarity brushes indicate an open equalizer (page 58). T____ F____

5. Brushes are all interchangeable if they fit the brush holder (page 59).
   T____ F____

6. Name two types of current flowing in a brush's conduction zone (page 60).

7. Split brushes reduce the circulating current in the conduction zone (page 61). T____ F____

8. A spiral groove cut in a commutator improves its performance (page 61).
   T____ F____

9. All brushes in a multiple brush holder should carry the same number of amperes (page 62). T___ F___

10. All motors should have their brush boxes the same distance from the commutator (page 62). T___ F___

11. Uneven distance between brushes can affect commutation (page 62). T___ F___

12. Spring tension must be the same for all brushes (page 63). T___ F___

13. High amperes require heavier spring tension (page 64). T___ F___

14. Brush shunt wire size isn't important because the brush box will carry the current (page 65). T___ F___

15. How can uneven commutator segments be detected (page 65)?

16. Name two causes of excessive brush dusting (page 65).

17. Gray brushes are better than black brushes for (page 66)
    a. high amperes.
    b. less wear.
    c. low amperes.

18. Give the formula for finding brush current density (page 66).

19. Silicone grease should be used on DC motors (pages 66–67). T___ F___

20. Why should brushes be properly seated (page 67)?

21. What will cause a brush face to have a flat black color (page 67)?

22. The final brush setting for a DC motor is the spot that has less than $1/2$ volt between segments (page 69). T___ F___

23. Moving the brush holder—one segment at a time against rotation—until there is the least amount of sparking is called the black band adjustment for a motor (page 69). T___ F___

24. Generators should always have their residual magnetism restored—by applying DC to F1 and F2—after AC has been used to set the brushes on neutral (page 69). T___ F___

25. Name two ways to test an armature on site (pages 70–71).

26. Name two symptoms of a grounded shunt field (pages 71–72).

27. All coils should be replaced when an ohmmeter reading is between 1 and 50 megohms (pages 72–73). T___ F___

28. A circuit that is being tested for a ground should be isolated from all other circuits (page 73). T___ F___

29. Why are shorted turns in the series field circuit usually very obvious (pages 76–77)?

30. A compound wound DC motor will run if the series field is not connected (page 78). T____ F____

31. How is a compound wound generator affected when the series field is not used (page 78)?

32. The commutator is a high-leakage circuit (page 79). T____ F____

33. Why are interpoles subject to problems if the armature is overheated (page 80)?

34. Why should AC and a resistor be used to supply AC for testing (page 82)?

35. DC will cause high current to flow in shorted turns (page 83). T____ F____

36. Why is it important to draw a schematic of a complicated circuit before testing it (page 84)?

37. Name four problems that occur in the shunt field (page 85).

38. An ohmmeter will show a lower resistance reading when the test is close to a grounded coil (page 88). T____ F____

39. Why is an equipment ground not recommended for DC machines (page 88)?

40. A coil with shorted turns will show a _____ reading than the rest of the coils (page 89).
    a. higher resistance
    b. lower resistance

41. Dividing the test voltage by the number of coils being tested will give the approximate voltage found across each coil—if none are shorted (page 89). T____ F____

42. A + or – of _____ difference is allowable when comparison testing coils with an ohmmeter (page 89).
    a. 2%
    b. 5%
    c. 10%

43. A + or – of _____ difference is allowable when comparison testing coils with a voltmeter (page 89).
    a. 2%
    b. 5%
    c. 10%

44. What will happen to the armature if the shunt field opens while it's operating with no load (page 90)?

45. Loose coils can lead to a grounded or open circuit (page 91). T___ F___

46. An open coil in a two-parallel circuit shunt field will _____ of the shunt field (pages 91–92).
    a. lower the resistance
    b. raise the resistance

47. An insulation breakdown between the shunt and series fields is usually caused by high amperes in the series field (page 94). T___ F___

48. Running a compound DC machine without the series field is a quick way of ruling out the series field as the problem (page 94). T___ F___

49. The resistance of the series field has very little effect on the amperes of its circuit (page 94). T___ F___

50. A grounded series field coil may not show a problem (page 95). T___ F___

51. The shunt field should never have AC voltage applied to it, because the series field will be destroyed by the high voltage being transformed into it (page 97). T___ F___

52. Leads F1 and F2 are usually larger than the rest of the DC machine's leads (page 99). T___ F___

Chapter 3

Electric Motor Theory

Nearly all commercial and domestic power consumed in the United States is AC. High-voltage AC power is carried great distances by power lines, with minimal power loss. Very little power is lost transforming AC to the voltage value required by a consumer. AC is much more versatile and practical than DC power. AC induction motors require much less maintenance than DC motors because most have no commutator. The variable-hertz drive gives the three-phase induction motor speed control that is almost as good as that of the DC motor.

AC Terminology

Generating Alternating Current

The operation of most AC motors is affected by the way AC power is produced. For this reason, the generating of AC power will be explained first.

AC alternators (also called generators) consist of a DC-excited rotor and a stator winding. Figure 3.1 shows a single-phase alternator. The stator has one winding or phase.

In an AC alternator (Fig. 3.2), the conductors of the stator poles are cut by magnetic lines of force (flux) provided by rotating DC field coils. For clarity, a magnet will be used in the following explanation. The magnet will travel past two poles of the stator. As the magnet travels past the first stator pole, its magnetic flux cuts the conductors of the pole's coils, producing a voltage.

In Fig. 3.3 the magnet is in the neutral position between the poles. As the magnet advances, its flux cuts the conductors of each stator pole, and voltage

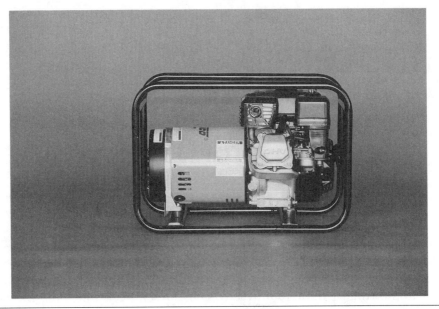

FIGURE 3.1 A single-phase alternator. *Winco, Inc.*

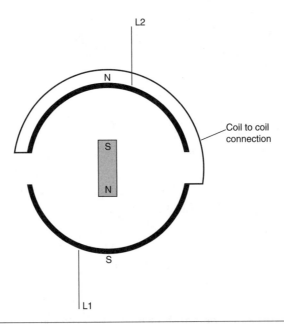

FIGURE 3.2 Basic components for generating alternating current (a bar magnet and a two-pole field).

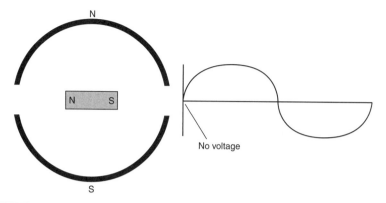

FIGURE 3.3   The bar magnet in the neutral position, where no voltage is generated.

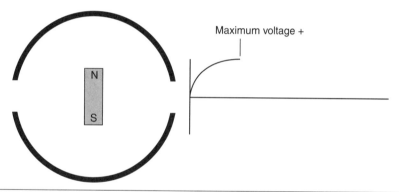

FIGURE 3.4   The magnet's position when maximum voltage is produced.

is produced in them. The voltage (output) value will increase (Fig. 3.4) until the magnet becomes aligned with the stator poles. The voltage is now at peak positive value. As the magnet moves out of alignment, the voltage decreases. When it again reaches the neutral position, the voltage is zero.

When the magnetic flux cuts the next pole (Fig. 3.5), the voltage output is negative. When the magnet becomes aligned (Fig. 3.6) with the stator poles, the output voltage is at its highest negative value. As the magnet moves out of alignment, the voltage output decreases until it is again in the neutral position. (One cycle of power has been produced.)

Cycles and Hertz

The number of cycles produced in 1 second is the power's frequency. If one cycle of power is produced in 1/60 of a second, the frequency of the power

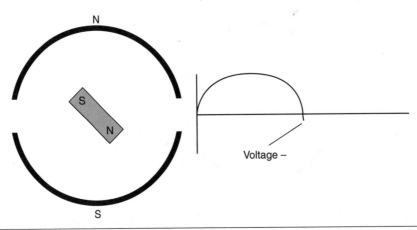

FIGURE 3.5   As the magnet enters the area of the opposite polarity, the voltage produced changes polarity.

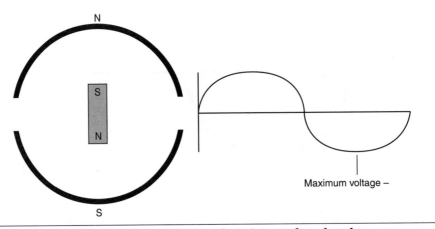

FIGURE 3.6   Maximum voltage (of this polarity) is produced at this spot.

is 60 cycles per second (60 Hz). The word *hertz* (Hz) is used instead of the phrase *cycles per second*.

Electrical Degrees and Time

Electrical degrees are used as a reference in electrical machines. Electric motors and alternators both have their poles arranged in a circle. The poles are placed symmetrically and precisely in the circle. Each pole equals 180 electrical degrees.

Electrical degrees are used to describe a location on a pole. The center of a pole is 90° from the edge of the first coil of a pole (Fig. 3.7). This reference

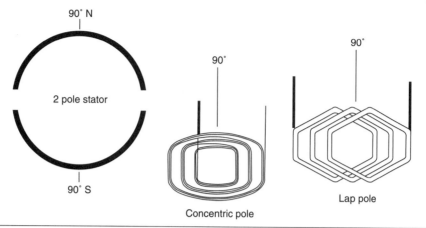

FIGURE 3.7   The center of a pole is the 90 *electrical degree* spot.

point is where a single-phase motor's start-winding pole is placed in relation to the run-winding pole (Fig. 3.8). Three-phase machines separate the phases 120° apart (Fig. 3.9).

One cycle of alternating current is produced by passing a magnet over two (180°) poles of a phase. One cycle is equal to 360 electrical degrees. The time taken to produce one cycle is also expressed in degrees (Fig. 3.10). Electrical degrees are used to specify the exact time (in relation to the voltage) that something occurs during a cycle. For example, if amperes and voltage are out of step (explained under "Inductive Reactance," later in the

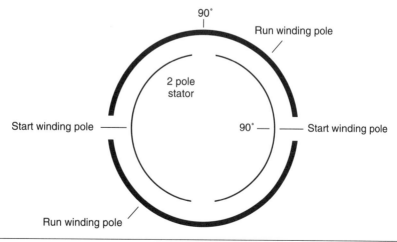

FIGURE 3.8   The start-winding coils are located on both sides of the 90° spot of the run-winding pole.

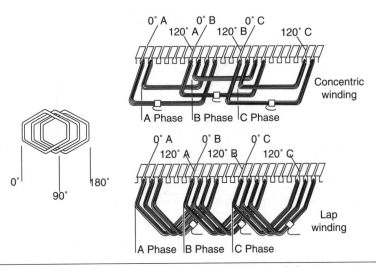

FIGURE 3.9 Three-phase poles are located 120 *electrical degrees* apart.

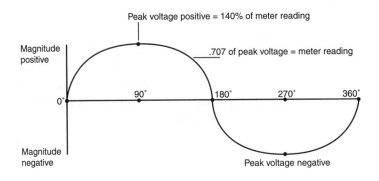

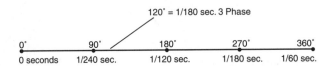

FIGURE 3.10 The sine wave includes voltage magnitude and the time it takes to produce it.

chapter), electrical degrees are used to describe the length of time that they are separated.

The word *angle* is used to describe the amount of separation (in time). An example is phase angle (phase means winding and angle means time).

Electrical Degrees and Mechanical Degrees

Two stator poles produce one cycle of power. Each pole contains 180 electrical degrees. A two-pole stator has the same number of electrical degrees as mechanical degrees. In a four-pole stator, there are 720 electrical degrees in the 360 mechanical degrees of the stator (Fig. 3.11).

Degrees and Location

Degrees are also used to reference a location on a pole. An example of this is the position of the start and run windings of a single-phase stator (Fig. 3.12). The 90° spot of a run-winding pole is the center of the pole (Fig. 3.7). The start-winding coils are located on either side of this spot (Fig. 3.8). Three single-phase windings of a three-phase motor are located 120 electrical degrees apart (Fig. 3.9).

Two-Phase Power

A two-phase alternator is shown in Fig. 3.13. There are two identical single-phase windings located 90 electrical degrees apart. When a DC field is rotated in this stator, two out-of-step voltages are produced. The two voltages are separated by 90 electrical degrees in time (1/240 second at 60 Hz).

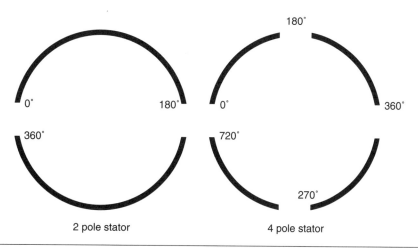

FIGURE 3.11   The two-pole stator has 360 mechanical degrees and 360 electrical degrees. The four-pole stator has 360 mechanical degrees and 720 electrical degrees.

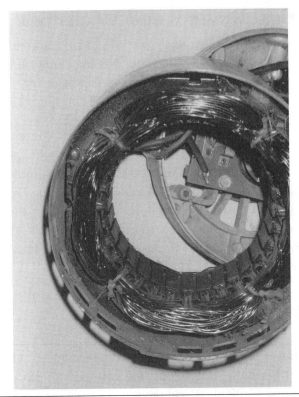

FIGURE 3.12 The stator of a four-pole single-phase motor. *Marathon Electric.*

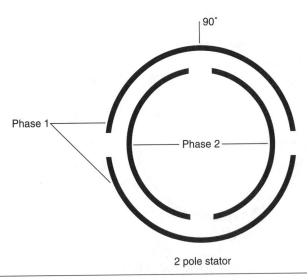

FIGURE 3.13 A two-phase motor has two identical phases spaced 90 electrical degrees apart.

Two-Phase Rotating Magnetic Field

When 60-Hz two-phase voltage is applied to the windings of a two-phase motor, the peak magnetic strength of its two windings will occur 1/240 second apart. This creates a rotating magnetic field in the stator.

The motor's rotor will try to rotate at the same speed as the two-phase alternator's DC field. Varying the speed of the alternator will vary the speed of the motor.

The rotating magnetic field of an AC motor is created when the current flows in each phase at a different time. When the time (in electrical degrees) that current flows in each phase matches the location (in electrical degrees) of the poles in each phase, the rotating magnetic field is at its effective best.

Figure 3.14 shows a two-pole, two-phase stator with a compass in its bore. Energizing phase 1 with low-voltage DC will cause the compass to align as shown in Fig. 3.14. Energizing phase 2 with the same polarity moves the compass needle 90° as shown in Fig. 3.15. Reverse the DC polarity and apply it to phase 1. The compass moves another 90° as shown in Fig. 3.16. Apply the reversed polarity to phase 2 and the compass needle aligns as shown in Fig. 3.17. Reverse the DC polarity and apply it to phase 1, and the needle will have made a complete revolution. This is a simplified version of the rotating magnetic field that occurs when two-phase voltage is applied to a two-phase winding.

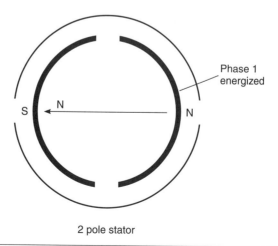

2 pole stator

FIGURE 3.14   **A two-phase stator with a compass needle (aligned with energized phase 1).**

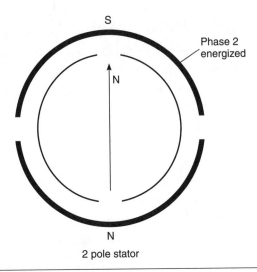

FIGURE 3.15   When energized, the compass needle moves 90 electrical degrees and aligns with phase 2.

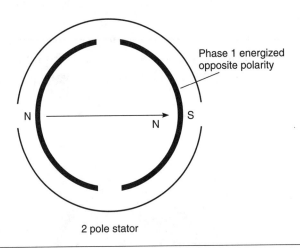

FIGURE 3.16   Reversing the polarity and applying it to phase 1 aligns the compass needle as shown.

Revolutions per Minute, Poles, and Hertz

The cycles per second (Hz output) is determined by the speed at which the AC alternator is driven and its number of poles.

The synchronous RPM of most AC motors is determined by the number of poles and the Hz of the power source. The following formulas give the synchronous speed of AC induction and synchronous motors:

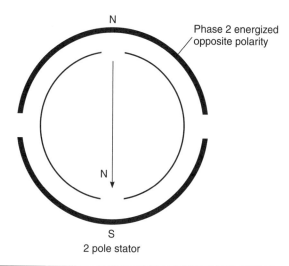

FIGURE 3.17   Applying the reversed power to phase 2 completes one revolution.

Hz × 60 seconds (to get cycles per minute) × two poles = numerator
Numerator ÷ poles = RPM
Numerator ÷ RPM = poles
60-Hz numerator = 7200

A two-pole motor on 60 Hz:

7200 ÷ 2 poles = 3600 RPM

Two-pole synchronous speed (at any Hz) divided by the motor's pairs of poles = synchronous speed of the motor:

For 60 Hz: 3600 ÷ pairs of poles = RPM
For 50 Hz: 3000 ÷ pairs of poles = RPM

The Single-Phase Alternator

Single-phase power is generated when a DC magnet is rotated past a single-phase winding (Fig. 3.18). As the magnet starts across a single-phase pole, the voltage generated builds to its peak value when the magnet and the pole are aligned. The voltage value drops to zero as the magnet goes past center and exits the pole. As the magnet enters the next pole, the voltage generated is of the opposite polarity. The voltage value will increase to a peak value and then decrease to zero.

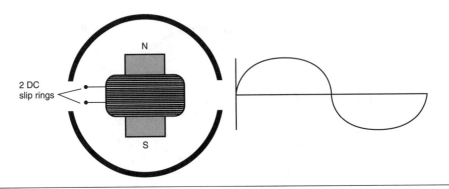

FIGURE 3.18 Basic principle of generating single-phase power.

Current, Ampere, Power, and Reactive Terms

Many combinations of terms are used with current amperes and power. Current, amps, and amperes are used interchangeably. Current is used to describe the fact that amperes of some type (lagging, leading, in unity with voltage) are flowing in a circuit. Ampere flow (or amps flowing) also signifies current in a circuit.

The current exchange between a coil and the power source may be called magnetizing current, magnetizing amperes, reactive current, reactive amperes, and reactive amps. The current between a capacitor and the power source may be referred to as reactive current, leading current, leading amperes, leading amps, reactive amperes, and reactive amps.

The amperes required to do work, or which are part of a wattmeter's result, are called true amps, true amperes, real amps, and real amperes.

Power terms include apparent power, real power, power factor, lagging or leading power factor, and actual power. All power includes both amperes and volts.

The term *reactive* can apply to both leading (capacitive) and lagging (inductive) descriptions.

The following text describes the effect different components have on the amperes of a circuit and the condition of a circuit's power.

Inductive Reactance

Inductive reactance is a form of resistance to current flow. Like DC current, alternating current creates a magnetic field around a wire. Alternating current is continually changing in magnitude and polarity. The strength of the magnetic field around a wire changes *with* the amount of current. The direction of current flow in the wire changes *with* the polarity of the AC voltage. The polarity of the magnetic field around a wire changes *with* the polarity of the AC voltage.

When two wires carrying AC current in the same direction are placed next to each other (Fig. 3.19a), there is a bucking interaction between them until their magnetic fields combine into a larger magnetic field (Fig. 3.19b). This interaction will cause the current to flow out of step with and later than the voltage that drives it. This current delay is called *inductance*. Resistance to current flow caused by inductance is called *inductive reactance*.

Inductive reactance occurs when there is a changing magnetism. Inductive reactance can be demonstrated with a length of wire. If the wire is in a straight line, the only significant resistance to AC current flow is the resistance of the wire itself. If the wire is wound into a coil, there will be inductance and inductive reactance (resistance to current flow). If the wire is wound around a loop of iron, as shown in Fig. 3.20, inductive reactance is increased even more.

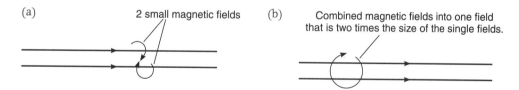

(a) 2 small magnetic fields (b) Combined magnetic fields into one field that is two times the size of the single fields.

FIGURE 3.19   (a) A magnetic field surrounds each current-carrying conductor. The fields buck each other while the current magnitude is changing. (b) They combine and form a stronger field.

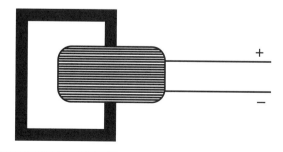

FIGURE 3.20   Iron adds to the bucking action of a coil.

Impedance

Impedance is the sum of the resistance of the conductor and inductive reactance. Inductive reactance furnishes most of the resistance to current flow in an AC motor.

Inductive Reactance and Counter-voltage

Inductive reactance in the windings of an AC motor serves the same purpose as counter-voltage in the armature of a DC motor. The resistance of the windings of a DC armature and the windings of an AC motor are very low. If DC voltage is applied to an AC winding, there would be no inductive reactance and the winding would be destroyed.

Power Factor

Power factor is defined as the ratio between true power and apparent power. True power is the power that is actually used by a load. Apparent power is the sum of the true power amperes and the magnetizing amperes (see the upcoming section "Magnetizing Amperes and Lagging Power Factor"). The power factor of a circuit is found by dividing true power by apparent power.

True power amperes can be found with a wattmeter. The meter's reading in watts divided by the applied voltage equals the true amperes.

Amperes, read with an AC ammeter, are the circuit's real power plus its apparent power. A clamp-on AC ammeter displays the sum of the magnetizing amperes plus the true amperes of an inductive circuit. True power amperes of a wattmeter divided by apparent power amperes of an AC ammeter equals the power factor. The result is a decimal value or percentage.

A power factor meter can be used to find the power factor of a circuit. It measures both the voltage and amperes, and computes the result.

The power supplier's watthour meter records and charges for only the real power.

Unity Power Factor

If there is no reactive power, the current and the voltage are in step (Fig. 3.21). The condition is called *unity power factor*. All the power used at unity power factor is real power and is recorded by the power meter as consumed power. (Power used by resistors or stove elements is at unity power factor.)

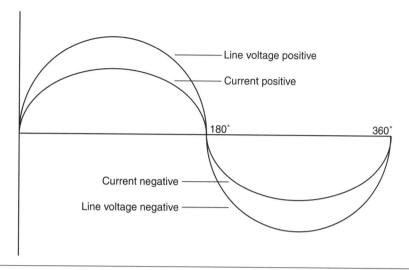

Line voltage positive

Current positive

180°

360°

Current negative

Line voltage negative

FIGURE 3.21   Current and voltage are in step when there is unity power factor.

Magnetizing Amperes and Lagging Power Factor

When amperes are out of step with the voltage, the circuit has a power factor. The power factor can be leading or lagging. (Leading power factor is covered under "Capacitor Function," later in the chapter.) This section explains a lagging power factor and the relationship of current and magnetism in a coil during one cycle.

As current flow increases from point A to point B in Fig. 3.22 through a coil of wire, there is a bucking interaction between the magnetic fields of each turn of wire. The magnetic fields of the entire coil then combine to form a large magnetic field. The power source supplies the current needed to magnetize the coil (magnetizing amps) until peak current (point B) is reached. The current at this point starts to drop toward zero. As the current decreases, the magnetic field starts to collapse. The collapsing magnetic lines of force cut the turns of the coil and create a voltage in them. The power this voltage creates makes current (or magnetizing amps) flow back to the power source. The current cycle now crosses the zero line (point C) and flows in the opposite direction. The coil is now being magnetized in the opposite polarity. It will draw power and then send it back to the source in the same sequence as the previous half-cycle. With inductance in the circuit, current is out of step with (*lagging*) the voltage by a number of degrees (which represent time). (The circuit has a lagging power factor as shown in Fig. 3.23.)

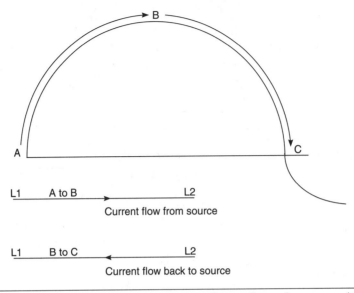

L1   A to B   L2

Current flow from source

L1   B to C   L2

Current flow back to source

FIGURE 3.22   As current increases, power is drawn from the source (A to B). When current decreases, power (B to C) is created and flows back to the source.

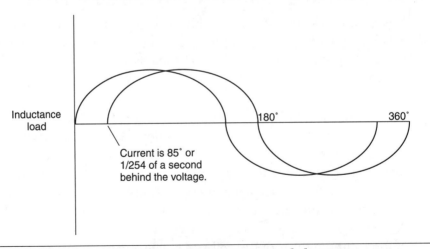

Inductance load

180°   360°

Current is 85° or 1/254 of a second behind the voltage.

FIGURE 3.23   The result of Figure 3.22 is continuously lagging current, or power factor.

The magnetizing amperes do no work. The more inductive components the circuit has, the higher the magnetizing amperes become. In a circuit with a low power factor, the magnetizing amperes may be more than the real amperes.

Power Factor in Motors

Most motors have a lagging power factor because they have coils surrounded by iron. Magnetizing a coil surrounded by iron creates much more inductive reactance than magnetizing a coil alone. The amount of resistance created by inductive reactance is determined by the number of turns of wire and the amount of iron surrounding the coil.

A motor's power factor can change under certain conditions. If the power factor is given on a motor's nameplate, it has been calculated with rated voltage, and with rated load applied. A motor's power factor is lowered if the applied voltage is higher than its nameplate rating or if the motor is underloaded. One or both of these conditions are normal in industry.

Low Power Factor Concerns

The problem created by a low power factor occurs in the supply lines and the power source. Magnetizing current (reactive or magnetizing amps) is added to the load current (true amperes) value (Fig. 3.24). The result is heat loss and voltage drop in the supply lines. Magnetizing amperes add nothing to the power cost and do no work. However, they do require the same number of circular mils (circular mils were explained in Chapter 1) as the real power amperes.

A transformer large enough to carry a load at unity power factor would overheat if the load's power factor were too low. A 50 percent power factor requires a transformer two times larger to avoid overheating (with the same load). Power suppliers often charge a penalty if they must provide a larger transformer or if they have to install power factor correcting capacitors. The effect of power factor correcting capacitors is explained later under "Capacitor Function."

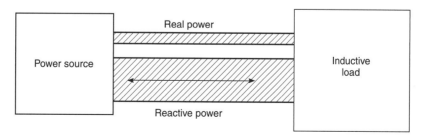

FIGURE 3.24   The difference between reactive power and real power.

Inductive Reactance and Hertz Change

The amount of resistance created by inductive reactance is related to the amount of polarity change (Hz) of the power source. As Hz go up, so does the resistance of the inductive circuit. The winding of a motor designed for 50 Hz has more resistance on 60-Hz power. With fewer amperes through its winding, the motor has less power than it had with 50 Hz. (As Hz go up, amps go down.)

The power change is proportional to the Hz change. If voltage is *increased* in the same proportion as the Hz increase, the lost power is restored.

Formula for 50-Hz to 60-Hz Power Change

There are two ways to increase a 50-Hz motor's power when it is operated on 60 Hz. One is to rewind the motor using fewer turns. The other is to increase the applied voltage. Both methods increase the volts per turn, which increases the amperes. The following formulas are used to increase the applied voltage:

$$\sqrt{60 \div 50} = 1.095 \quad \text{Increase the voltage 10\%}$$
$$60 \div 50 = 1.2 \quad \text{The square root of } 1.2 = 1.095$$
$$1.10 \times \text{the 50-Hz voltage} = \text{the same horsepower but less torque}$$
$$60 \div 50 = 1.2$$
$$1.2 \times \text{the 50-Hz voltage} = \text{the same torque but more horsepower}$$

The increased horsepower will increase the motor's heat, but the increase in RPM will increase the motor's cooling ability. In most cases, the increased cooling compensates for the increased heat.

Impedance in a Three-Phase Winding

There are two forms of resistance to current flow in AC motor windings: resistance of the wire and inductive reactance of the winding. The sum of the resistance and inductive reactance is called *impedance*.

The following is a hypothetical example of three-phase motor data. It will show the amount of resistance to current flow that inductive reactance has in the total impedance of the motor.

- Hypothetical motor nameplate data: 230 V, 5 A
- Winding data: connection one delta (the delta connection puts each phase across the line).

- Four-pole: three coils per pole group, 25 turns of #17 wire; each turn contains 1 ft of wire. (#17 wire has 5.05 ohms of resistance per 1000 ft.)
- Impedance of one phase of a four-pole three-phase motor is as follows:

Resistance:

3 coils × 25 turns (1 ft per turn) = 75 ft per pole group
75 ft × 4 pole groups = 300 ft of 17 wire per phase
      300 ft ÷ 1000 ft = 0.3 (multiplier to get resistance of 300 ft)
    0.3 × 5.05 ohms = 1.515 ohms (resistance of one phase)

Inductive reactance:

230 volts ÷ 5 amperes = 46 ohms of impedance per phase
46 ohms (impedance) – 1.515 ohms of resistance =
                    44.48 ohms of inductive reactance

If 230 volts DC are applied to one phase:

            Volts ÷ ohms = amps
230 volts ÷ 1.515 ohms = 151.8 amps (through one 17 wire)
1.515 ohms ÷ 46 ohms = 3.2%

Only 3.2 percent of the resistance to current flow is furnished by the resistance of the winding. Inductive reactance furnishes the remaining 96.8 percent of the resistance.

Capacitance and Capacitive Reactance

Capacitance is the term used to describe a component's or circuit's ability to store electrical energy.

Capacitive reactance is the amount of effect capacitance has on a circuit. Like inductive reactance, capacitive reactance has resistance to current flow.

Capacitor Function

A capacitor stores power in electrostatic form. As a fully charged capacitor is discharged, power is produced. Unlike inductive reactance, as Hz go up, amperes also go up.

The sine wave in Fig. 3.25 shows how the electrons flow in a capacitor. As the voltage goes up at point A, electrons are pulled into the capacitor plate as if there is a vacuum. When the voltage reaches peak value (B), the electron (current) flow stops because the voltage potential on the capacitor

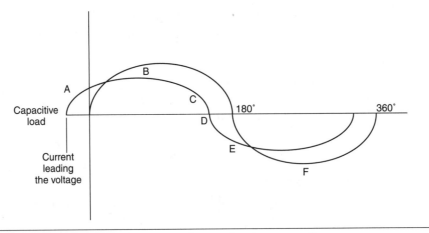

FIGURE 3.25   The changing current's reaction to a capacitor.

plate equals the voltage of the sine wave. As the voltage value drops (C), the electrons are expelled from the capacitor plate as though they are under pressure. At point D in the sine wave, there is no voltage potential, so there is no electron flow. As the voltage polarity reverses (E), the electrons are pulled onto the opposite plate as though there is a vacuum. At the peak of this half of the cycle (F), electron movement stops. As the voltage drops toward zero, the electrons leave the plate (as though under pressure) until the voltage drops to zero. The vacuum-pressure action results in a separation of the voltage and the current, causing the current to lead the voltage. *This is a line condition with a leading power factor. (The current leads the voltage.)*

Capacitor Value

The capacitance rating of a capacitor is measured in microfarads (mfd or μ). *(μ is used instead of mfd on the side of capacitors.)* A capacitor's rating is affected by the area of its plates and the thickness of its dielectric (insulation between plates).

Increasing the plate area increases the mfd rating. When capacitors are connected in parallel, the plate area is increased. The capacitor's mfd ratings are added together to get the total capacitance value in a parallel connection.

When two identical capacitors are connected in series, the thickness of their dielectric is doubled. This connection reduces the total capacitance value. To get the capacitance value of identical capacitors connected in series, divide the value of one capacitor by the number of them. For example, for two 300-mfd capacitors connected in series, 300 mfd ÷ 2 (capacitors) = 150 mfd.

Capacitors for Single-Phase Motors

There are many types of capacitors. The two types discussed here are used in single-phase motors—oil-filled and electrolytic capacitors (Fig. 3.26). Both have two plates made of aluminum. The aluminum plates are separated by insulation called a dielectric. The plates and dielectric are rolled up tightly and placed in a container.

Oil-Filled Capacitor

Oil-Filled Capacitor in Single-Phase Motors

The oil-filled capacitor is used to improve the power factor of capacitor-start, capacitor-run motors. In a permanent-split capacitor-run motor it will shift

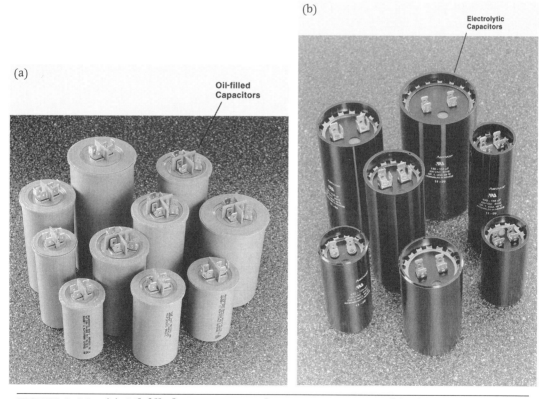

FIGURE 3.26   (a) Oil-filled capacitors used in motors. *Aerovox.* (b) Electrolytic capacitors used in motors. *Aerovox.*

the current flow of the start winding ahead (in time) of the current flow in the run winding. The result is a rotating magnetic field.

Oil-Filled Capacitor Components

The oil-filled capacitor has two aluminum plates that are separated by a thin sheet of plastic. The aluminum plates are very thin because this type of capacitor limits the amperes to a low value.

Aluminum is melted and sprayed on the plastic dielectric to form the plates. This results in a rough surface, which increases the plate area for a given length of dielectric. This design reduces the overall size of the capacitor.

The physical size of the oil-filled capacitor is very large compared to the electrolytic capacitor. This is because the dielectric of an oil-filled capacitor is much thicker than the dielectric of an electrolytic capacitor.

The oil-filled capacitor is designed for continuous service. The oil is used to cool the capacitor. The plates and dielectric are rolled tightly, insulated, and sealed in a metal container filled with oil.

The voltage rating of oil-filled capacitors should be at least twice that of the motor.

Oil-Filled Capacitor Connection in a Two-Value Capacitor Single-Phase Motor

The oil-filled capacitor is used to improve the power factor of a single-phase motor. It's connected in series with the start winding (Fig. 3.27). They are always connected in parallel with the start switch contacts and the electrolytic capacitor. If two or more are used, they are always connected in parallel with each other and are never connected in series.

The start contacts are open when the motor is running. The electrolytic capacitor is now disconnected. There is now a circuit from line 1—through the oil-filled capacitor and the start winding—to line 2. The amperes through the start-winding circuit are leading the voltage. This reduces the magnetizing amperes caused by the run winding's low power factor. The number of leading start-circuit amperes is subtracted from the amperes of the run winding. Figure 3.28 shows the amperes of each circuit, with the result displayed on the line ammeter. If the capacitance is doubled, as in Fig. 3.29, the start-winding circuit amperes are doubled. The run-winding amperes stay the same, and the line amperes are correspondingly less.

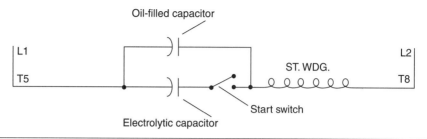

FIGURE 3.27   The oil-filled capacitor is connected in series with the start winding.

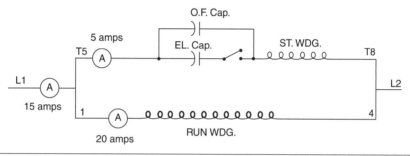

FIGURE 3.28   The effect that an oil-filled capacitor has on amperes of a motor.

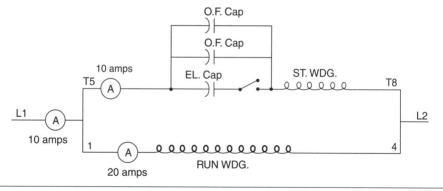

FIGURE 3.29   Line amperes (magnetizing amps) go down as more capacitance is added.

The ampere values in the preceding illustrations are exaggerated. In reality, the start winding would get hot and char if amperes were too high for its wire size.

Mounting the Oil-Filled Capacitor

Use *caution* when mounting the oil-filled capacitor. Be sure the lid is free to move.

When the dielectric of oil-filled capacitors breaks down, an arc forms between the plates. The arc vaporizes the oil and causes the container to swell, rupture, and possibly catch fire. The lid of the container is designed so that it automatically disconnects the capacitor, therefore eliminating the problem.

The lid contains the external terminals, as shown in Fig. 3.30. When it's in place, a protrusion under the terminals presses against the lead straps of the plates. The lid moves upward when vaporized oil has caused the container to expand. The terminals are then automatically disconnected from the plates.

Electrolytic Capacitors

Electrolytic Capacitor Function in the Capacitor-Start Motor

Figure 3.31 shows the motor circuitry in the start position with the start switch contacts closed. When the motor is energized, it will accelerate to approximately 75 to 80 percent of synchronous speed in about 1 second. At

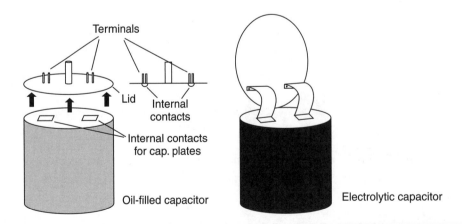

FIGURE 3.30   The oil-filled capacitor disconnects itself when the lid is forced away from its internal contacts. The electrolytic capacitor isn't sealed and has straps riveted to the lid terminals.

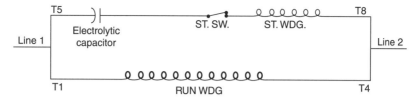

FIGURE 3.31 The "at rest" position of a capacitor-start motor's start contacts.

this time, the switch contacts open, shutting off the start-winding circuit. The motor now operates with only the run or main winding energized.

Electrolytic capacitors (of the correct size) help create a rotating magnetic field that starts the motor. They control when current flows in the start-winding circuit (in relation to current flow in the run winding). The sine wave in the following illustrations shows why the capacitor has to be the right size.

In Fig. 3.32, there are no capacitors in series with the start winding. The sine wave shows the position of the line voltage, the start-winding amperes, and the run-winding amperes.

Run windings and start windings both have inductance. The run winding usually has more turns and is located deeper in the iron (slot) than the start winding. Both factors (more turns and deeper in the iron) give the run winding more inductive reactance than the start winding. More inductive reactance in the run circuit causes the amperes to flow slightly later than the amperes of the start circuit. This small difference in time between the start and run currents means the motor will have very little starting torque.

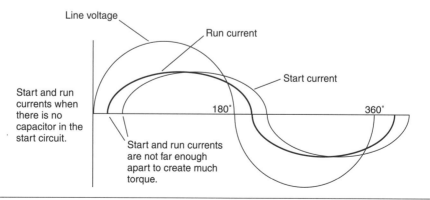

FIGURE 3.32 The currents of the start and run windings have little separation in time with no capacitor in the circuit. Very little torque is produced.

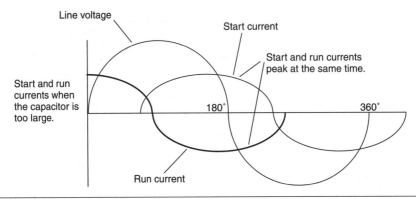

FIGURE 3.33   The effect is the same with too much capacitance as with no capacitance.

Figure 3.33 is an example of too much capacitance. In this illustration, the two currents are flowing at nearly the same time. The result is little or no starting torque.

Figure 3.34 shows an ideal amount of separation. The separation is 90 electrical degrees or 1/240 second on 60-Hz power. Ninety degrees separation in the current matches the 90 physical degrees of separation between the start and run windings. This timing is accomplished using capacitor(s) of the right size in series with the start winding.

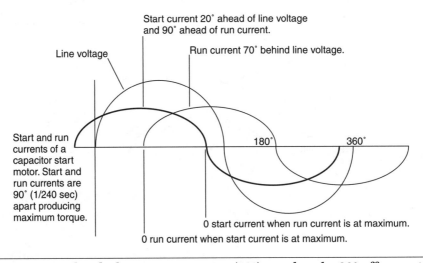

FIGURE 3.34   The ideal current separation (90°) matches the 90° offset position of the start and run windings.

Electrolytic Capacitor Components

The plate area of the electrolytic capacitor is small compared to the oil-filled capacitor. The aluminum plates are thick because the electrolytic capacitors are designed for high current. Paper impregnated with a solution of water, borate, and glycol separates the plates. The liquid solution is an electrolyte which conducts electrons to one of the plates during each half cycle.

The dielectric of the electrolytic capacitor is the very thin oxidized coating on the aluminum plates. The extremely thin dielectric gives this type of capacitor a very high mfd rating for its size.

The components are encased in a plastic container with the terminals mounted in the lid. Straps of aluminum connected to the plates are riveted to the terminals. A small hole (sealed with rubber) is also in the lid. The rubber seal will rupture and release the pressure (when the capacitor's water boils from overheating).

 Caution   Prolonged current flow (more than 3 seconds) will make the water solution boil and can cause the electrolytic capacitor to explode.

Electrolytic Capacitor and Start-Winding Connections

Electrolytic capacitors are connected in series with the start switch and the start winding as shown in Fig. 3.35. There is no standard sequence for connecting the components of this circuit.

Multiple electrolytic capacitors are usually connected as a single unit in the connection sequence. They are connected either parallel or in series with each other when more than one capacitor is used.

Parallel Electrolytic Capacitor Connection

The parallel connection is shown in Fig. 3.36. The total capacitor plate area is increased with this connection. The mfd value of the capacitors is added to get the total capacitance of the circuit. The mfd rating of the capacitors connected in parallel can be different.

FIGURE 3.35   The capacitor-start motor's internal connections.

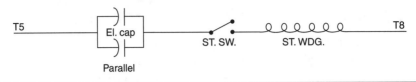

FIGURE 3.36 Parallel connected capacitors.

The capacitor's voltage rating must be as high or higher than the voltage applied to the start-winding circuit.

Capacitors are connected in parallel for two reasons: If a large mfd value is needed, two or more capacitors are easier to mount on the motor than one large one, and it enlarges the capacitor's cooling area.

Heat develops in a capacitor each time the motor starts. Frequent starting may prevent the capacitor from cooling sufficiently. All capacitors have a breakdown voltage value. When voltage reaches this value, it will puncture the capacitor's dielectric and destroy it. As the temperature of the capacitor increases, it takes less voltage to break down the dielectric.

If frequent starting is necessary, a single capacitor can be replaced with two or more capacitors in parallel. (The total value of the parallel capacitors must be the same as the single capacitor's value.) Two or more capacitors in parallel increase the cooling area, allowing more starts per hour.

The highest recommended temperature for capacitors is 150°F. High temperatures dry out the electrolyte fluid. Loss of electrolyte fluid derates the mfd value of a capacitor.

Series Electrolytic Capacitor Connection

The series capacitor connection is used on high-voltage (above 200 volts) motors. When two low-voltage capacitors are connected in series, their dielectric is doubled, allowing the voltage to be doubled. The cooling area of two low-voltage capacitors is larger than that of one high-voltage capacitor.

There are two rules for connecting electrolytic capacitors in series:

- Never have more than two capacitors connected in series with each other.
- When two capacitors are connected in series, they must be labeled the same mfd value.

The reason capacitors must be of the same mfd value is that the voltage divides across them in inverse proportion to their rating. A lower mfd capacitor would have voltage across it that is too high.

Example:

$$\text{Volts} \times C_1 \div C_1 + C_2 = \text{volts across } C_2$$
$$\text{Line voltage} = 240 \text{ volts}$$
$$\text{Capacitor \#1 } (C_1) = 200 \text{ mfd}$$
$$\text{Capacitor \#2 } (C_2) = 100 \text{ mfd}$$

240V $\times$ C$_1$ $\div$ C$_1$ + C$_2$ = 240V $\times$ 200V $\div$ (C$_1$) 200 + (C$_2$) 100 or,
48,000 $\div$ 300 = 160 volts across C$_2$ (the smaller capacitor rated for 120V)

Under normal starting conditions, the voltage developed across the capacitor- and start-winding circuit will be 140 percent of the applied voltage for less than one-half of a cycle. Capacitors of equal mfd rating would have (half of 240 volts) 120 volts $\times$ 140 percent = 168 volts across each of them. The voltage across C$_2$ in the above example would be 160 $\times$ 140 percent = 224 volts.

The actual mfd rating of the electrolytic starting capacitor is within 10 percent of its stamped rating. If there are two ratings, the actual mfd value will be somewhere in between. Such a small mfd difference isn't critical because it's normally in the circuit for 1 second or less.

Testing the Electrolytic Capacitor

A capacitor's rating can be checked with an electronic test instrument like the one shown in Fig. 3.37. This instrument can quickly give a capacitor's value in mfd. The test uses low energy and is not destructive.

FIGURE 3.37   An instrument for checking a capacitor. *EXTECH Instruments.*

An ohmmeter can be used to test a capacitor. When testing a good capacitor, it will show a low-resistance reading, which rises slowly until the capacitor is charged.

This test indicates the capacitor has capacitance but doesn't give the amount of capacitance. No reading means the capacitor is open; a steady low-resistance reading indicates it's shorted.

Some internal problems in the electrolytic capacitor aren't revealed by using a low-energy test method. A test method (formula to follow) using line voltage will stress the internal components and determine the capacitor's mfd value. This test will also cause a weak part to break down. (A capacitor has nearly this same amount of stress each time the motor starts.)

One of a capacitor's weak spots is under the lid where the lead straps of the plates are riveted to the connection terminals. These straps flex when the capacitor heats and cools, causing them to crack or break at the rivet. The motor occasionally fails to start when this happens. The high amperes of this test method will blow the weak spot open.

Excessive heat dries out the water-based electrolyte in the capacitor. (Too many start cycles per hour is a common cause of overheating.) A thick piece of insulation is needed between the capacitor and the hot motor frame to limit the heat transfer.

A capacitor's mfd rating is lowered when some of the electrolyte evaporates. This condition lowers the capacitor's amperes.

Capacitor Test Formula

The following formula determines the capacitor's exact mfd rating:

[159,300 ÷ Hz] × [amps ÷ volts] = mfd

The formula can be shortened to a single number when the Hz value is constant. This number divided by the test bench voltage results in a multiplier. The multiplier times the capacitor's (across the line) amperes value equals the capacitor's mfd rating:

$$159,300 \div 60 \text{ Hz} = 2655$$
$$2655 \times [\text{amps} \div \text{volts}] = \text{mfd}$$
$$2655 \div \text{volts} = \text{multiplier}$$
$$\text{Multiplier} \times \text{capacitor amps} = \text{mfd}$$

Below are several test bench voltages and multipliers. Apply line voltage to the capacitor and quickly read the amperes. Multiply the resulting amperes and the multiplier for the capacitor's mfd value.

 Caution Place a box or pail over the capacitor during this test because the capacitor could explode and cause injury.

2655 ÷ 110 V = 24   24 x 35 amps = 840 mfd

2655 ÷ 120 V = 22   22 x 35 amps = 740 mfd

2655 ÷ 130 V = 20.4   20.4 x 35 amps = 714 mfd

2655 ÷ 230 V = 11.5   11.5 x 20 amps = 230 mfd

2655 ÷ 240 V = 11   11 x 20 amps = 220 mfd

Replace the capacitor if test results are 20 percent less than the capacitor's stamped rating.

If 50-Hz power is used, divide 3186 by the test bench voltage.

Determining the Right Size Capacitor

Figure 3.38 shows the connection used to find the right size capacitor for a capacitor-start motor. However, the sequence of the start-circuit components isn't always as shown in this diagram.

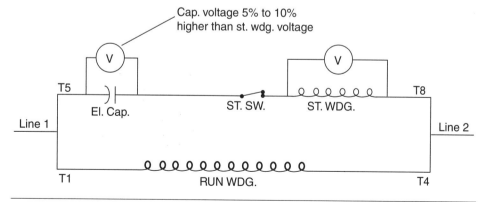

FIGURE 3.38   Schematic for determining the right size capacitor.

NEMA identification numbers for start-winding leads are T5 through T8.

The capacitor terminals are usually accessible, but the location of the start winding isn't always visible. An ohmmeter can be used to locate the start winding. Disconnect one of the capacitor leads to make sure it isn't in the circuit. Place one probe on the disconnected capacitor lead and the other on T5. If there's no reading, put the probe on the other capacitor lead. If there is a reading, turn the motor's shaft. If the start winding is between the probes, a high and low fluctuation will show on the ohmmeter. If there is no fluctuation, move the probes to the other capacitor terminal and T8, and turn the shaft.

Two voltmeters should be connected (as shown in Figure 3.38) so that the time power applied is kept to a minimum. Lock the rotor, apply the motor's rated voltage, quickly read the two voltmeters, and shut off the power. The voltage read across the capacitor should be 5 to 10 percent higher than the voltage across the start winding. If the voltage across the capacitor is too high, do the test again using a *smaller* size capacitor.

If the start winding is dual voltage, use the same procedure as just described, with the low-voltage multiplier. Figure 3.39 is a schematic of a dual-voltage start winding. The capacitors are always located between T6 and the start winding and between T8 and the start winding. Both capacitors must have the same mfd rating.

The switch contacts will be as shown or as a separate circuit located between T9 and T10.

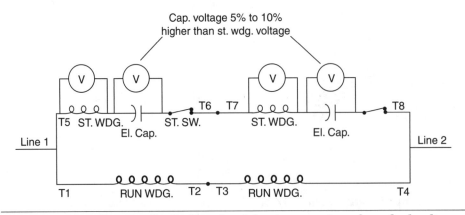

FIGURE 3.39 Schematic for finding the right size capacitors for a dual-voltage start winding.

If the original capacitor isn't available or readable, use the following data to select a capacitor size that's close enough to start the process:

400 mfd ÷ hp @ 120 volts (dual-voltage motors use a 120-volt value)
100 mfd ÷ hp @ 240 volts

The preceding values should not be used to select the actual size capacitor for a motor. Do not allow the start winding to overheat. Electrolytic capacitors must be of the same mfd value when connected in series with each other.

Operation of the Single-Phase Capacitor-Start Motor

The capacitor-start motor starts the same as a two-phase motor. But, unlike two-phase power, which has two voltages, there is only one voltage applied. Single-phase power is applied to both the start and run windings to start the motor. About 1 second after power is applied, the rotor should be turning between 75 and 80 percent of synchronous speed. At this time a switch (Fig. 3.40) disconnects the start winding.

The motor now runs on only the run (main) winding. With rated load applied, the rotor pulls the load at or near its nameplate RPM. With no load,

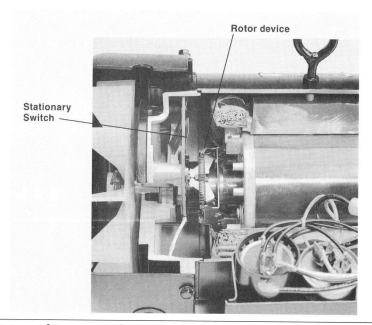

FIGURE 3.40 The start-winding controls. *Marathon Electric.*

the rotor will turn at nearly synchronous RPM. (The load determines the exact speed of the motor.)

The capacitor-start and three-phase induction motors both have squirrel cage rotors. To better understand how the rotor gets its power, transformer theory must first be understood.

Transformer Theory

A transformer is a device used to transfer the power of alternating current from one circuit to another without a direct electrical connection. Figure 3.41 is a schematic of a transformer.

Transformer Components

A transformer consists of a primary winding, a secondary winding, and a laminated iron core.

The primary winding is connected to an AC power source. It transforms power to the secondary winding. The secondary winding is on the load side of the device. The laminated iron carries the magnetic flux of the primary winding to the secondary winding.

Voltage applied to the primary winding creates a continually changing current in its circuit. The changing current creates a magnetic field in the iron core, which is changing at the same rate. The laminated core's magnetic flux cuts the conductors of the secondary winding, transforming a voltage into it. The only connection between the two windings is the magnetic link provided by the laminated iron.

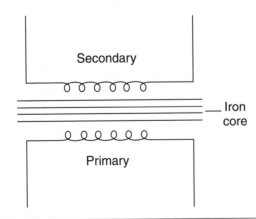

FIGURE 3.41 The schematic of a transformer.

Transformer Function

Assuming there are no losses, the same voltage value of each turn of the primary winding is transformed into each turn of the secondary winding. In the design of most transformers, 5 to 8 percent more turns are added to the secondary winding to account for any losses.

Volts Per Turn

The primary coil (Fig. 3.42) has 100 turns with 100 volts applied to it: 100 volts ÷ 100 turns of the primary coil = 1 volt per turn in the primary coil. Each turn of the secondary coil will have 1 volt transformed into it. The voltage value of each turn in the secondary is added to the next turn because they are wound in the same direction. If there are 100 turns in the secondary coil, its voltage output is 100 volts: 10 turns = 10 volts, 1 turn = 1 volt, 200 turns = 200 volts, and so on.

Transformer Rating

Power for the load is furnished by the primary winding when a load is applied to the secondary winding. Although the voltage and amperes of the

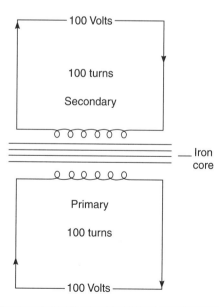

FIGURE 3.42   The volts per turn in the primary determine the volts per turn in the secondary.

primary and secondary may vary, the watts—assuming there are no losses—are the same in both windings. The watt load of the primary is determined by the load applied to the secondary.

The physical size (amount of iron and wire size) of a transformer is determined by its power rating. Transformers and motors are similar in this respect. In the United States, transformers are rated in VA or KVA (volts × amps or volts × amps ÷ 1000) and electric motors are rated in horsepower. In Europe, both transformers and electric motors are rated in voltamperes or kilovoltamperes (VA or KVA).

For example (with no losses):

Primary load is 1000 watts @ 100 volts
Secondary load is 1000 watts @ 10 volts
Primary winding: 1000 watts ÷ 100 volts =
　　　　　　　　　　　　　　10 amperes in the primary winding
Secondary winding: 1000 watts ÷ 10 volts =
　　　　　　　　　　　　　　100 amperes in the secondary winding

Transformers are designed to step the voltage up or down, or to isolate the power, with no change in voltage value. Hertz is not changed by a transformer.

Solder Gun Transformer

A solder gun makes a good model for transformer theory. The power values and problems are similar to those of the squirrel cage rotor of an induction motor.

Figure 3.43 shows a solder gun and its components, a primary winding, a secondary winding, and laminated iron (which forms a magnetic link between the two windings). The solder gun data is 120 volts and 300 watts, and has 260 turns in the primary coil and one turn in the secondary one.

The formula for volts per turn is

Volts ÷ turns = volts per turn

The formulas for watts are

Volts × amps = watts
Watts ÷ amps = volts
Watts ÷ volts = amps

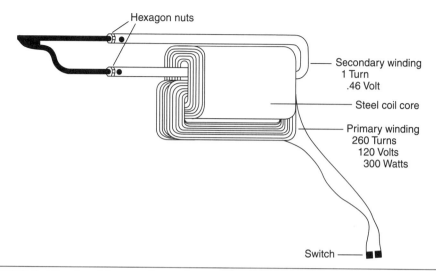

Hexagon nuts

Secondary winding
1 Turn
.46 Volt

Steel coil core

Primary winding
260 Turns
120 Volts
300 Watts

Switch

FIGURE 3.43   A solder gun has voltage values that are close to those of a squirrel cage rotor.

Primary volts per turn and amperes formulas are

120 volts ÷ 260 turns = 0.46 volt per turn (primary)

300 watts ÷ 120 volts = 2.5 amperes (primary)

The formulas for secondary volts per turn and amperes are

1 turn = 0.46 volt

300 watts ÷ 0.46 volt = 652 amps

0.46 volt ÷ 652 amps = 0.0007 ohm (resistance of the soldering tip)

The Squirrel Cage Rotor

Components of the squirrel cage rotor are rotor bars, end rings, and laminated iron.

The bars and end rings of the squirrel cage rotor are illustrated in Fig. 3.44. The bars (located in the slots of the rotor iron) are all connected together by end rings. Rotor bars are made in many shapes and sizes, as shown in Fig. 3.45.

End rings join all the rotor bars. The size of the end rings depends on the number of bars covered by the face of one pole. Their size must be slightly larger than the total circular mil area of one-half the bars, covered by the face

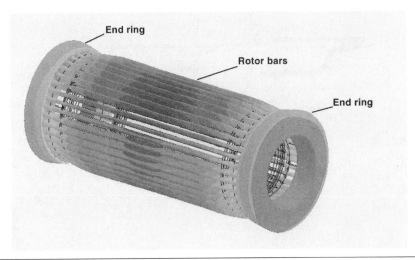

FIGURE 3.44 The squirrel cage winding without iron surrounding the winding. *Siemens.*

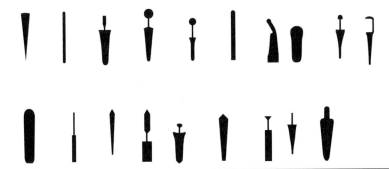

FIGURE 3.45 Some examples of the many rotor bar shapes.

of one pole. The end rings of a two-pole motor are much thicker than those of a four- or six-pole motor.

Rotor bars (and end rings) are made of many different materials. Most standard single- and three-phase induction motors use cast aluminum. Copper, brass, and alloy combinations are used in special-duty motors and some standard ones.

Laminated iron completes the magnetic circuit from pole to pole.

Current Flow in Stator Coils

Figure 3.46 shows the coils of a two-pole stator laid flat. The two coils are connected, so they have opposite polarity. The centers of the stator poles are

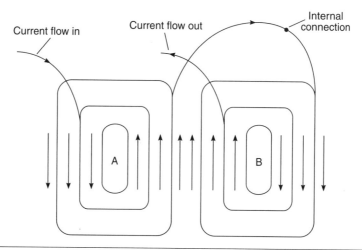

FIGURE 3.46   Two-pole stator coils showing the internal connection.

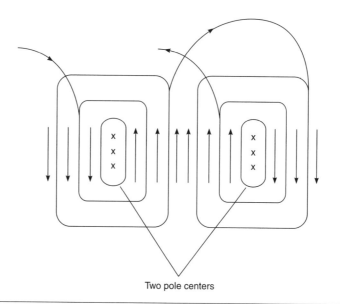

FIGURE 3.47   Electromagnets formed by current flow in the coils of a two-pole stator.

at points A and B. The center coils of the stator poles (Fig. 3.47) will straddle at least two slots. Current flow in the center coil of the poles is in opposite directions on both sides of the straddled slots. (This creates a pole center like that of a DC shunt field pole.)

Current flow in the outer coils of each stator pole is in the same direction. Because the current flows in the same direction, the outer coils may be located in adjacent slots, or they may share the same slot.

All stator coils are well insulated (from the slot iron). The thickness of the insulation is determined by the operating voltage value.

Function of the Squirrel Cage Rotor

The voltage transformed into the bars of the squirrel cage winding from the stator coils is very low. Rotor bars don't have to be insulated from the iron—unlike stator coils—because the voltage transformed into them is so low. The turn ratio of the stator winding and the squirrel cage winding steps the voltage down to less than 1 volt (similar to the solder gun).

Current flow in the rotor bars creates poles that are the same shape as the stator poles (coil groups). Figure 3.48 shows the current flow in the bars of a squirrel cage rotor at loaded speed.

Stator pole groups are connected so that their polarity is opposite the one next to it. This polarity sequence becomes the same in the rotor. The rotor bars will each have a different amount of power transformed into them. The amount of power is related to the position of the bars in reference to the stator poles.

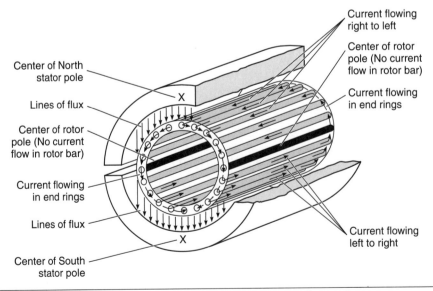

FIGURE 3.48   Current flow in the rotor bars creates poles in the rotor iron.

The bars located at the center (90°) of the stator poles are cut by the most lines of force; therefore, they create the most power. Bars located away from center (90°) create progressively less power. When a bar's angle to the pole face is less than 90 electrical degrees, fewer lines of force cut it. Bars located in the neutral area don't create any power. No lines of force are cutting them at this position. (The iron around the idle bars becomes the rotor's pole centers.)

The opposite polarity stator pole transforms power of the opposite polarity into the bars located within its pole face. Current flows through the end rings from bars of one polarity to bars of the opposite polarity. Two poles form in the rotor iron (which is centered in the neutral area of the stator poles). A repel and attract magnetic force is established between the stator and rotor windings. This magnetic interaction creates torque that pulls the load.

Rotor Hertz and Amperes

The induction motor is very comparable to a transformer. The primary amperes of a transformer are determined by the load applied to the secondary. If the secondary winding were shorted, the primary winding would overheat and be destroyed.

In an induction motor, the rotor bars (secondary winding) are shorted together by end rings. At 0 RPM (locked rotor) the rotating magnetic field of the stator is at synchronous speed. The magnetic lines of force are cutting the rotor bars at maximum Hz (cycles per second). This transforms a very high AC current into the rotor bars. (The stator is furnishing this power, so the amperes are also very high in the stator.) As the rotor accelerates, the difference between the synchronous speed of the stator's rotating magnetic field and the speed of the rotor decreases. Rotor Hz become lower as rotor speed increases. As rotor Hz go down, fewer lines of force cut its rotor bars, resulting in fewer amperes in both the rotor and stator. With rated load applied, the rotor speed will be nameplate RPM, and stator amperes will be at nameplate value. At loaded speed, rotor Hz are about 1.5 to 3 cycles per second. The amperes are high enough in the rotor winding (at this RPM) to produce the motor's rated torque.

Slip, Hertz, and Motor Speed

The difference between synchronous speed and nameplate speed is called *slip*. Slip produces a calculated number of amperes, torque, and horsepower that equal the motor's rating. A motor rated continuous duty won't overheat at rated RPM.

The following calculations start with the basics and explain the relationship of poles, power supply Hz, synchronous speed, slip, and rotor Hz:

One pole = 180 electrical degrees.

Two poles = 360° (the number of degrees in one cycle).

One Hz of power is produced when a magnet passes one pair of poles in 1 second. Sixty Hz = 60 pairs of poles passed in 1 second.

Hz × 60 seconds = pairs of poles passed in 1 minute.

60 Hz × 60 seconds = 3600 pairs of poles passed in 1 minute.

3600 ÷ number of pairs of poles = synchronous RPM @ 60 Hz.

3600 × 2 = 7200 (the total number of poles passed in one minute).

7200 ÷ number of poles = synchronous RPM @ 60 Hz.

7200 ÷ synchronous RPM = number of poles in a 60-Hz motor.

Any Hz × 60 seconds (to get cycles per minute) × 2 = numerator.

Numerator ÷ poles = synchronous RPM.

Numerator ÷ synchronous RPM = poles.

Slip = the difference between synchronous speed and loaded speed.

1800 – 1750 (nameplate RPM) = 50 RPM of slip.

50 ÷ 1800 × 100 = 2.7% slip.

Hz × % of slip = rotor Hz.

60 × 2.7% = 1.66 rotor Hz.

Synchronous speed of a four-pole motor on 60 Hz is 1800 RPM. High-slip rotors have 5 percent slip or 3 rotor Hz. Synchronous motors have no slip. Their poles will be misaligned a few degrees (behind the stator poles). The amount of misalignment is determined by the load and the strength of the rotating field.

Synchronous Speed and Zero Torque

When there is no load, a squirrel cage rotor will accelerate to near-synchronous speed. The rotor Hz are 0, with no torque at synchronous speed. Figure 3.49 illustrates why.

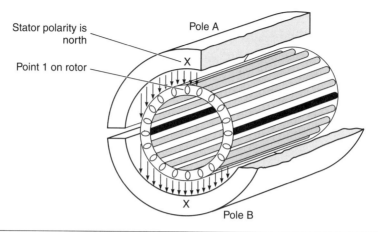

FIGURE 3.49   No torque is produced by the rotor windings at synchronous speed because no lines of force are cutting the rotor bars.

Stator pole A (in Fig. 3.49) is north and point 1 on the rotor is aligned with it. By the time point 1 of the rotor aligns with pole B of the stator, the polarity of pole B has changed to north. (This compares to the rotor being stationary in a DC magnetic field.)

At synchronous speed, no lines of force cut the rotor bars and no power is transformed into them. When no current flows in the rotor bars, there is no torque.

If the rotor was to somehow accelerate faster than synchronous speed, it would generate power instead of producing torque. The rotor iron is magnetized, and its flux cuts the stator coils, producing power.

Rotor Bar Stress

The squirrel cage winding of the rotor has a wide range of amperes. Rotor amperes are very high every time the motor starts. This causes the rotor bars to get hot. The expansion rate of the rotor bars is much greater than that of the iron. The resulting stress can break the rotor bars.

There are many magnetic polarity changes in the rotor iron from the time a motor starts until it reaches loaded speed. This creates constant vibration in the rotor bars. As a rotor bar crosses each stator slot and aligns with the iron of the next stator tooth, there's a different magnetic pull.

Rotor bars are skewed to reduce the slot- and tooth-caused vibration. They will crack or break if they aren't tight in the slot.

Cast aluminum squirrel cage windings are used in most standard induction motors. Casting the aluminum rotor bars and end rings results in a tight fit. This reduces the chance of bars getting loose and cracking from vibration.

Aluminum bars and end rings must be grossly oversized to prevent the normally high starting current from melting them. Oversized bars reduce the amount of magnetic path of the rotor. This is why rotors with cast aluminum windings are larger than rotors with windings made of copper or other alloys.

Broken Rotor Bars

Broken rotor bars decrease a motor's power. Current flowing in each bar contributes to the motor's torque. The current flow is all but stopped when a bar is cracked or broken. The cracked bar is embedded in iron, but because the voltage is so low, very little current is carried by the iron.

The tip of a solder gun is a good example of low voltage and high current. When the solder gun has had a lot of use, the tip doesn't get hot when the gun is turned on. Even if the nuts holding the tip seem very tight, tightening them slightly makes it work as it should. (Very low resistance virtually stops current flow when voltage is this low.)

The number of broken rotor bars and the load determine whether or not the motor fails. The motor's amperes are much lower than normal (with no load). (Amperes in the stator winding depend on the power demand of the rotor winding.) A maximum-loaded motor with broken rotor bars will soon fail. The stator windings will look charred—as though they failed—from being overloaded.

Cracked End Rings

Cracked end rings cause uneven torque and decrease the motor's power, resulting in failure. The failed winding—as with open rotor bars—will be charred as though the motor failed from being overloaded.

The size of the end rings is determined by the number of rotor bars a stator pole covers. End rings must be larger than the circular mil area of half of the rotor bars that are covered by the pole.

Aluminum end rings sometimes have a bubble or piece of slag embedded in them. Any type of flaw will reduce the circular mil area at this point. A reduction in circular mils restricts current flow and the spot gets hot.

High starting current will melt the aluminum at this spot, opening the end ring. When an open aligns with a neutral spot (between the stator poles),

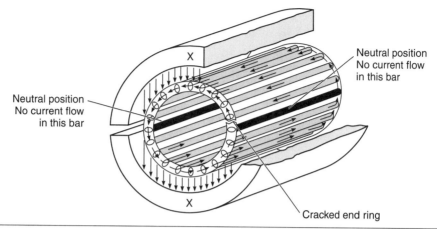

FIGURE 3.50   A cracked end ring stops current to flow when it is a neutral position. All current has to flow through the opposite neutral location in the end ring.

the entire rotor current is forced to flow, as shown in Fig. 3.50. The current then doubles in the part of the ring located in the adjacent neutral spot. High current will soon melt the end ring in this area, creating another open.

Misaligned Rotor Iron

When the rotor iron is misaligned with the stator iron, the size of the motor's magnetic path is decreased. The result is loss of torque and increased magnetizing amperes. The increased magnetizing amperes overheat the motor (even with no load).

Like a motor with broken rotor bars, a fully loaded motor soon fails. The windings will look charred—as with a motor that has failed from an overload.

Other than noticing high no-load amperes, the only way to identify this problem is to view it. It may be necessary to remove one end frame to see the misalignment.

Chapter 3 Review

1. Unlike the DC generator, which has a commutator and brushes, AC power is taken directly from the stator winding of the AC alternator (page 105). T___ F___

2. The stator of a single-phase alternator has one winding (or phase) (page 105). T___ F___

3. The changing position of the exciter magnet (as it sweeps by poles of an alternator's stator) produces the voltage variation of a cycle (page 107).
   T___ F___

4. Frequency of power is expressed in cycles per minute (page 107).
   T___ F___

5. Electrical degrees are used to reference both time (during a cycle) and the location (on a pole) because electric machines have a circular shape (page 109). T___ F___

6. A single-phase alternator has one phase; a single-phase motor has a start phase and run phase (page 111). T___ F___

7. Electrical degrees and mechanical degrees are the same in a two-pole motor (page 111). T___ F___

8. The three windings of a three-phase motor are _____ from each other (page 111).

9. Two phases (start phase and run phase) are symmetrically offset when they are located 90° from each other (page 111). T___ F___

10. The most efficient torque is produced when the current flow in two windings is offset (degrees in time) the same as the windings are offset (degrees in location) (page 113). T___ F___

11. The Hz value of an AC induction motor's power supply determines its approximate RPM (pages 114–115). T___ F___

12. Current flow can be out of step (in time) with voltage (page 116).
    T___ F___

13. Inductive reactance is a form of resistance (page 117). T___ F___

14. Reactive power always describes the current lagging the voltage (page 117).
    T___ F___

15. All current-carrying conductors have a magnetic field around them (page 117). T___ F___

16. The power of a magnetic field _____ when current increases (page 117).
    a. increases
    b. decreases

17. A bucking action between magnetic fields of two AC current-carrying conductors creates resistance to (page 117)
    a. voltage.
    b. current flow.

18. The bucking action (described in #17) _____ as Hz increase (page 117).
    a. increases
    b. decreases

19. The resistance (caused by inductive reactance) increases with the length of a wire (in a straight line) (page 117). T___ F___

20. Impedance is the total resistance of an inductive circuit (page 118). T___ F___

21. Power suppliers charge for only the true amperes of a circuit (page 118). T___ F___

22. A wattmeter can be used to find true amperes of an inductive circuit (page 118). T___ F___

23. An ammeter can be used to find true amperes of an inductive circuit (page 118). T___ F___

24. The power factor that is given on the nameplate of a motor is an accurate value for all conditions (page 121). T___ F___

25. Two problems associated with low power factor are voltage drop and an added penalty cost on the power bill (page 119). T___ F___

26. A motor designed for 50 Hz will have the same horsepower as 60 Hz if the voltage is decreased proportionately (page 122). T___ F___

27. The motor described in "Impedance in a Three-Phase Winding" can be tested using 230 volts DC (pages 122–123). T___ F___

28. A capacitor will cause the current to (page 124)
    a. lead the voltage.
    b. lag the voltage.

29. Oil-filled capacitors lower the two-value capacitor motor's amperes, also lowering the power bill proportionately (page 126). T___ F___

30. An oil-filled capacitor can be in the circuit (page 126)
    a. 1 second.
    b. 1/2 hour.
    c. continuously.

31. The electrolytic capacitor can be in the circuit (page 128)
    a. 1 second.
    b. 1/2 hour.
    c. continuously.

32. The capacitor-start motor starts on two-phase current and operates on single-phase voltage (pages 129–130). T____ F____

33. Multiple capacitors in parallel can be used to replace a single capacitor as long as the total mfd is the same (page 132). T____ F____

34. Why must capacitors that are connected in series have the same mfd rating (pages 132–133)?

35. An ohmmeter can be used to find a capacitor's mfd rating (page 134). T____ F____

36. An ohmmeter is used to determine if a capacitor (page 134)
    a. is open.
    b. is shorted.
    c. has capacitance.
    d. all of the above.

37. An electrolytic capacitor never becomes weak (page 134). T____ F____

38. The line voltage test can destroy a capacitor, but finds flaws that some electronic instruments miss (page 134). T____ F____

39. A motor's capacitor is selected according to horsepower rating alone (page 137). T____ F____

40. The start winding of a capacitor-start motor is in the circuit (page 136)
    a. less than 2 seconds.
    b. 5 seconds.
    c. continuously.

41. The capacitor-start motor will run at exactly nameplate RPM from no load to full load (pages 137–138). T____ F____

42. The primary winding of a single-phase transformer is (page 138)
    a. connected to the secondary.
    b. linked to the secondary by magnetic lines of force.
    c. the load side of the device.

43. The amperes of the primary are determined by the load applied to the secondary (pages 139–140). T____ F____

44. A solder gun transformer and the squirrel cage winding of a rotor have similar voltage and ampere values (page 140). T____ F____

45. The squirrel cage winding forms poles similar to stator poles (page 144). T____ F____

46. Power in the squirrel cage winding (page 145)
    a. comes directly from the line.
    b. flows through the shaft to the bars.
    c. is transformed from the stator windings.

47. When the rotor is locked, the rotor Hz is the same as the stator Hz (page 145). T____ F____

48. Amperes go down as rotor speed increases (page 145). T____ F____

49. The load determines the speed of the squirrel cage induction motor (page 145). T____ F____

50. The squirrel cage winding produces no torque at synchronous speed (pages 146–147). T____ F____

51. Broken rotor bars decrease a motor's power (page 148). T____ F____

52. Why is current nearly eliminated in a cracked rotor bar (page 148)?

53. A motor that fails from broken rotor bars and/or cracked end rings will look like a motor that had failed from being overloaded (page 148). T____ F____

54. Misaligned rotor iron will lower the motor's amperes (page 149). T____ F____

Chapter 4

The Single-Phase Motor

Single-phase motors are used where electric motor horsepower demands are small, such as in homes, office buildings, ranches, and farms. Although single-phase motors will operate on any two of the three wires of three-phase power, their use on three-phase power is discouraged by power suppliers. A three-phase load should be balanced as close as possible among its three conductors. A single-phase motor load is hard to balance because it isn't a constant load; an example of a constant load is lighting.

Three kinds of single-phase induction motors discussed in this chapter are *shaded-pole, split-phase*, and *capacitor types* (capacitor-run, capacitor-start, and two-value capacitor).

Shaded-Pole Motor

The shaded-pole motor is an inexpensive motor that is used on low-torque applications. Its size range is $1/2$ hp or less. The only moving part is a high-slip squirrel cage rotor. These motors are replaced, not repaired.

The start winding (shaded-pole winding) is a copper band surrounding about a third of the pole face (of each pole). When the motor is energized, a voltage/current flow is transformed into the copper band. This current flow offsets the time of peak magnetism (in the iron it surrounds) from the rest of the magnetism on the pole face. The result is an inefficient rotating magnetic field. The current flowing continuously in the shaded winding results in this motor running hotter than most types of motors.

Multispeed Shaded-Pole Motor

Some shaded-pole motor windings have multiple taps, giving them multiple speeds (Fig. 4.1). Lower speed selections are accomplished by adding more turns or coils between line 1 and line 2. (This lowers the amperes and weakens the torque.) The rotor's high-slip design allows the load to establish a lower speed without overheating. With no load, the motor will accelerate to near synchronous speed on all speed selections.

It is very important to connect the common lead to one side of the line (Fig. 4.1). If high and low leads are connected across the line, the windings between them will burn instantly.

Troubleshooting the Shaded-Pole Motor

The shaded-pole motor will develop worn bearings after years of use. The bearings are usually made of porous bronze that is surrounded with an oil-impregnated wicking. Oil seeping through the porous bronze lubricates the bearings. When the porous bronze becomes plugged, the bearings seize and stall the motor. In some cases, thin oil (such as WD-40) will free the bearings. The motor may run in this condition until a replacement is found. When the motor is mounted vertically or at an angle, the lower bearing fails first.

Some shaded-pole motors have a thermal cutout bulb embedded in the stator slot, next to a coil. If there is no response when the motor is energized, the thermal unit could be open and unable to reset itself. Bypass the thermal bulb by soldering its leads together. (The motor should operate satisfactorily without it.)

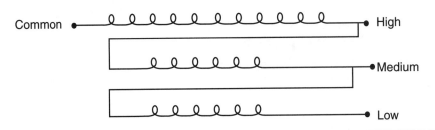

FIGURE 4.1   Schematic of the internal connections in a three-speed shaded-pole motor.

Split-Phase Motor

Split-phase motors are used in applications such as belt-driven furnace fans and appliances like washers and dryers. This motor has moderate to good torque. They range in size from fractional to 1 hp.

Components of the Split-Phase Motor

The main components of the split-phase motor are a stator, a start-winding switch assembly, and a rotor.

The stator has two windings, the run winding and the start winding (Fig 4.2). The start winding will have some type of device (switch assembly) that shuts it off after the motor starts.

The start-winding switch assembly in Fig. 4.3 has two components, a stationary switch mounted in the end frame and a centrifugal device

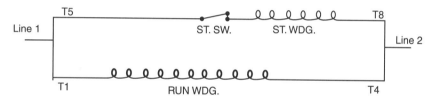

FIGURE 4.2   Schematic of a split-phase motor.

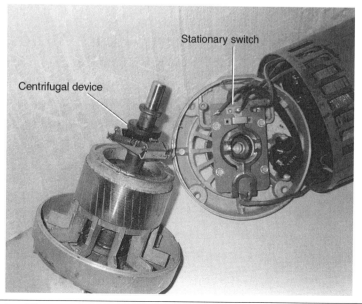

FIGURE 4.3   The stationary switch and centrifugal device of a split-phase motor.

mounted on the shaft of the rotor. The stationary switch has a set of contacts that controls the start winding.

The centrifugal device holds these contacts closed when the motor is at rest.

The rotor has a low-slip design like most capacitor-start and three-phase induction motors. The squirrel cage windings are usually made of cast aluminum.

Operation of the Split-Phase Motor

Both windings are energized when the motor starts. The start winding is made with small wire. The split-phase motor is sometimes called a resistance start motor. Because of its small wire size (high-resistance winding), the start-winding coils reach peak magnetism about 1/480 second before the run winding does. This is between 40 and 50° separation instead of the ideal 90° (1/240 second) separation the capacitor-start motor has.

When the split-phase motor starts, the weights on the centrifugal device pull the plastic spool away from the stationary switch contacts and shut off the start winding. The start winding is shut off at 75 to 80 percent of synchronous speed. The run winding brings the load up to nameplate RPM or above. Other start-winding control methods are discussed later in the chapter, under "Electronic Start Switch" and "Time-Delay Electronic Start Switch."

Troubleshooting the Split-Phase Motor

The split-phase motor components (except for the capacitor) are similar to those of the capacitor-start motor. Troubleshooting methods for the capacitor-start motor (described later) can be used on the split-phase motor. Split-phase motors are replaced rather than repaired when there is a problem.

Capacitor Motors

The types of capacitor motors are

- Capacitor-run
- Capacitor-start
- Two-value capacitor motor (capacitor-start, capacitor-run motor)

The Capacitor-Run Motor

Capacitor-run motors are used when high starting torque isn't required. They are quite trouble free because of their simple design. The high-slip squirrel cage rotor is its only moving part.

The high-slip rotor is designed for high inertia loading. An example of a high inertia load is the heavy squirrel cage fan in some furnaces. The fan is mounted directly on the shaft of the capacitor-run motor. The time involved in bringing the fan up to speed isn't a problem for the capacitor-run motor. This same type of load would seriously overheat the start winding of a split-phase motor.

Components of the Capacitor-Run Motor

The components of a capacitor-run motor are a high-slip squirrel cage rotor, an oil-filled capacitor, and a stator (with a start winding and run winding). Both start and run windings are high impedance windings, so they can start a high inertia load without overheating.

Operation of the Capacitor-Run Motor

When single-phase voltage is applied, the start and run winding are both energized. Because of the oil-filled capacitor, current flow peaks in the start winding before it peaks in the run winding, creating a rotating magnetic field. (The start and run windings are energized all the time.) These windings draw a small amount of starting current compared to a split-phase or capacitor-start motor. The start-winding current is limited by the oil-filled capacitor.

Although this motor normally runs hot, the electrical components are relatively trouble free. Its weakest components are the bearings. To be competitively manufactured, most of these motors have undersized sleeve bearings and a thin steel frame. If they are equipped with ball bearings, they tolerate angle positions better. Because of this motor's low cost, they are replaced rather than repaired.

Multispeed Capacitor-Run Motor

Some capacitor-run motors have multiple taps, giving them multiple speeds. Lower speed selections are accomplished by adding more turns or coils between the end of the high-speed winding and line 2 (Fig. 4.4). Extra turns lower the amperes and weaken the motor's torque. The rotor's high-slip design allows the load to lower the motor's speed without overheating it.

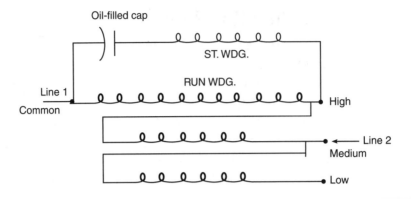

FIGURE 4.4   Schematic of a multispeed capacitor-run motor.

Without a load, the motor accelerates to near synchronous speed on all speed selections.

It is very important to connect the common lead to line 1 as shown in Fig. 4.4. Connecting leads high and low across the line would instantly burn the windings.

Reversible Capacitor-Run Motor

Reversible capacitor-run motors are used on low-torque loads such as house fans or dishwashers. Its schematic is shown in Fig. 4.5. Both start and run windings have the same winding data. An external switch places the oil-filled capacitor in series with one winding or the other. The winding in series with the capacitor becomes the start winding.

Troubleshooting the Capacitor-Run Motor

The most common problem with this motor is bearing failure. Some are equipped with porous bronze bearings that depend on oil seeping through the bronze for lubrication. When the porous bronze becomes plugged, the bearing will seize and stall the motor. In some cases, applying thin oil like WD-40 to the bearing keeps the motor running until a replacement is found.

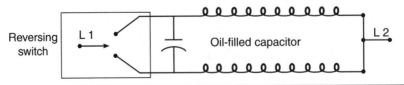

FIGURE 4.5   Schematic of a reversible capacitor-run motor.

When the motor is mounted vertically or at an angle, the lower bearing will fail first.

Winding failures involve the following:

- Burned start or run winding
- Open start or run winding
- Shorted start-winding or run-winding coil

A *burned start winding* is sometimes caused by a shorted oil-filled capacitor. The start-winding current is no longer limited by the capacitor, so it becomes extremely hot. A burned run winding can be the result of start-winding failure, bearing failure, or an overload. The motor must be replaced if the start or run windings are burned.

An *open start* or *run winding* may result from a breakdown in the insulation that separates them. Opens can also be caused by a breakdown in the slot insulation, shorting the winding to the stator core. A motor with this problem must be replaced.

Open lead-to-coil or coil-to-coil connections are less serious. This type of open can be located with an ohmmeter and a sharp needle probe. Fasten one test lead to a motor lead and test each connection by piercing the magnet wire. The open connection can be cleaned and soldered.

A number of things can cause a *shorted start-winding* or *run-winding coil*. Among them are contaminants, voltage spikes, damaged wire insulation, and overload. All of the shorted winding conditions require that the motor be replaced. The cause of the short should be remedied if it could affect the replacement motor.

Oil-filled capacitors will be either shorted or open when they fail.

An analog ohmmeter can be used to detect a shorted or open capacitor fault. If the capacitor is normal, the meter will peg toward low resistance and then slowly drop back and read the leakage value. If the capacitor is shorted, the ohmmeter will show no resistance. An open capacitor will show no reading or a steady, very low reading.

The Capacitor-Start Motor

The capacitor-start motor has good starting torque. It has a wide range of applications. Its size ranges from fractional to 35 hp.

Components of the Capacitor-Start Motor

The main components of the capacitor-start motor are a stator (with a start winding and run winding), a low-slip rotor, a start-winding switch assembly, and an electrolytic capacitor (Fig. 4.6).

Operation of the Capacitor-Start Motor

Single-phase voltage is applied to both start and run windings. An electrolytic capacitor in the start-winding circuit creates a leading current flow. The start-winding current leads the run winding current by 90° (or 1/240 second). The result is a very efficient rotating magnetic field and high starting torque. When the rotor reaches 75 to 80 percent of synchronous speed, its centrifugal device releases the stationary switch contacts and disconnects the start winding. The rotor accelerates the load to nameplate RPM or above.

Troubleshooting the Capacitor-Start Motor

Following are the most frequent component breakdowns of the capacitor-start motor, in descending order:

- Start switch assembly (stationary switch and centrifugal device)
- Electrolytic capacitor
- Start winding
- Run winding
- Bearings
- Rotor

Start Switch Assembly   The start switch assembly is responsible for connecting and disconnecting the start winding. The stationary switch contacts arc, vaporizing some of their material each time the motor starts. Over time, the contact area will no longer be large enough to make a good connection (resulting in dead spots, as described next). With the contact area

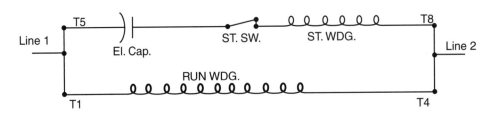

FIGURE 4.6   Schematic of a capacitor-start motor.

reduced, high starting current can melt the contacts, fusing them together (the start winding will then stay energized). The motor will be noisy and won't get up to full RPM. If the high-current protection device doesn't function, the start winding and capacitors overheat and fail.

Dead Spots   Worn contacts, centrifugal device spool, or thrust washers can cause the motor to have dead spots. It won't start unless the shaft is moved slightly. Moving the shaft changes the spool pressure point against the stationary switch, the contacts close, and the motor starts.

To verify that the switch is the component that is intermittently faulty, disconnect T5 and T8. Connect an ohmmeter across these leads and turn the shaft. If no open circuit occurs, pull the shaft away from the switch end of the motor and, at the same time, turn the shaft. (The stationary switch is normally located on the end opposite the shaft.)

Magnetic Center and Thrust Washers   Excessive shaft end play can cause intermittent starting problems. The rotor will try to seek its magnetic center. (The thrust washers should keep the rotor at this alignment.) If the rotor is not at magnetic center, the motor loses power and the magnetizing amperes are increased. This was discussed in Chapter 3 under "Misaligned Rotor Iron."

If the start contacts aren't closed when the rotor is aligned properly, the switch assembly should be adjusted accordingly.

Centrifugal Device Problems   If the preceding tests don't reveal an open circuit in the start winding, the problem may be in the rotating device or the capacitor (if it is in the capacitor, see "Open Capacitor" a bit later in the chapter). If the centrifugal device is dirty or worn, it can cause intermittent start problems. Cleaning or replacing it requires disassembling the motor or taking it to a repair center. When the centrifugal device is replaced, it must be positioned in exactly the same place as the old one.

The decision to replace the start switch assembly components depends on the overall condition of the motor and the repair cost. An electronic start switch may be the best option. Operation of the electronic start switch is explained later in the chapter under "Alternative Start-Winding Switches."

The motor's original stationary switch contacts and centrifugal device can be replaced with start switch devices, also described later under "Alternative Start-Winding Switches."

Normal Braking Effect Caused by the Electrolytic Capacitor   It's normal for a brief braking action to take place as a capacitor-start motor coasts to a stop. As the motor slows, the centrifugal device closes the start contacts. The capacitor will now discharge into the closed circuit of the start and run windings. The power coming from the capacitor is DC (Fig. 4.7). (When direct current flows in the stator windings of an induction motor, there is a strong braking effect on the rotor.)

Some three-phase motor controls use DC braking to slow a coasting type load. DC power can't, however, hold the motor's shaft stationary.

Electrolytic Capacitor   The following problems can occur with an electrolytic capacitor:

- Shorted
- Open
- Weak capacitor (loss of capacitance)

Shorted Capacitor   A motor with a shorted capacitor has very little starting torque. The peak magnetism time of its two windings is too close together (see "Operation of the Capacitor-Start Motor" earlier).

An ohmmeter can be used to test the capacitor, but (because of possible internal parallel circuits) it's necessary to disconnect one of the capacitor leads.

 Short the capacitor terminals together and discharge it before using the ohmmeter. Meters are available that can measure the capacitance of capacitors.

A normal capacitor will make an analog ohmmeter needle peg as if there is a short; then it will drop back slowly and read the capacitor's normal leakage. If the capacitor is shorted, the needle remains at the pegged position.

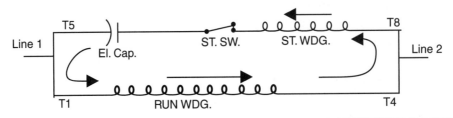

FIGURE 4.7   Current flow when the capacitor discharges into the motor circuit.

Open Capacitor   If a motor has only one capacitor, the start winding is disconnected when the capacitor is open and the motor won't start.

The capacitor can be tested for an open with an ohmmeter or a test light. One location in a capacitor where opens occur is under the lid. The connection straps for the plates are riveted to the terminals under the lid. These straps flex when the capacitor heats and cools, causing them to crack and break at the rivet. (The lid's retaining ring can be pried out to check this problem.) An open capacitor is sometimes an intermittent problem. The problem can be mistaken for dead spots caused by faulty start contacts.

When the stationary switch is replaced because of faulty contacts, the capacitor should be checked for cracked or broken straps.

Some centrifugal devices don't put even pressure on the stationary switch's wear pads. The contacts open and close (flutter) when the rotor approaches disconnect speed. Contact flutter will decrease contact life. It also produces voltage spikes in the start winding that will destroy the capacitor. Excessive contact flutter can cause frequent capacitor breakdown.

Weak Capacitor   Overheating is the main cause of a capacitor becoming weak. A capacitor is overheated if there are too many starts per hour or if the capacitor is mounted in direct contact with the motor shell. A weak capacitor reduces the motor's starting torque.

Overheating dries out some of the capacitor's electrolyte fluid. Loss of electrolyte fluid decreases the amperes available to the start winding and lowers the motor's torque. Applying line voltage and using the formula explained under "Capacitor Test Formula" in Chapter 3 will determine if the capacitor is weak.

If the motor doesn't have good starting torque, its capacitor may have been replaced with one that is too small. In this case use the method explained in Chapter 3 under "Determining the Right Size Capacitor."

Start Winding   All four of the following problems require that the motor be rewound or replaced:

- Burned
- Open
- Shorted coils
- Shorted to the run winding

Burned Start Winding   The start winding will overheat if it is energized for more than 3 seconds. An overload or a faulty start switch that doesn't allow the start contacts to open is frequently the cause of burned start windings. (Burned windings can be seen and/or smelled.) A motor with burned or discolored start coils must be rewound or replaced.

Open Start Winding   An open in the start winding can occur if it is shorted to the run winding. Because of the start winding's smaller wire size, it will melt in two. Most motor manufacturers don't put insulation between the start and run windings the way electric motor repair centers do. Rewound motors are less prone to start winding to run winding shorts.

It's common for opens to occur in motors wound with aluminum magnet wire. If the environment is corrosive (for example, a washing machine), a small bare spot in the wire will corrode it in two. Electric motor repair centers never use aluminum wire.

If the start winding is not burned or discolored, the open can be located with a test light or an ohmmeter. For example, a faulty internal lead-to-coil or coil-to-coil connection.

The following procedure should be used: Disassemble the motor and remove the insulation from the lead-to-coil and coil-to-coil connections. Start at one end of the start winding and proceed toward the other end as shown in Fig. 4.8. An open connection can be scraped clean and resoldered.

Smaller capacitor-start motors with a faulty start winding should be replaced. The larger capacitor-start motors can be economically rebuilt if the cost is near that of a new motor. Service centers use a higher temperature class insulation than is in the original factory-installed winding. Most use class H (Fig. 4.9). Service centers also insulate between the start and run windings.

Shorted Coils—Dual-Voltage Start Winding   Some capacitor-start motors have dual-voltage start windings (Fig. 4.10). The dual-voltage start winding has two pairs of lead wires numbered T5 through T8. Half the start-winding circuit is between numbers T5 and T6, and the other half is between T7 and T8. Capacitor(s) are always located (as shown in Fig. 4.10) between the start winding and T6 and between the start winding and T8. The start switch contacts will be located as shown in Fig. 4.11 or 4.12.

The two circuits have identical data, so they can be comparison tested. It is unlikely that both circuits would have the same problem. A microhmmeter or limited AC current can be used for this test. The current shouldn't be so high that it overheats the start-winding coils.

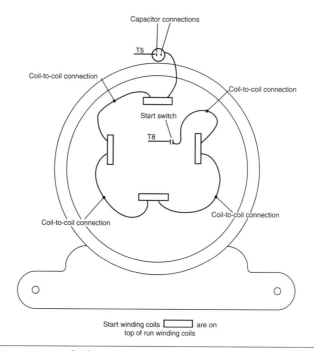

FIGURE 4.8   Locating a faulty connection in the start winding.

Insulation Class	Temperature Rating
A	105 degrees C
B	130 degrees C
F	155 degrees C
H	180 degrees C

FIGURE 4.9   Temperature chart for insulation classification.

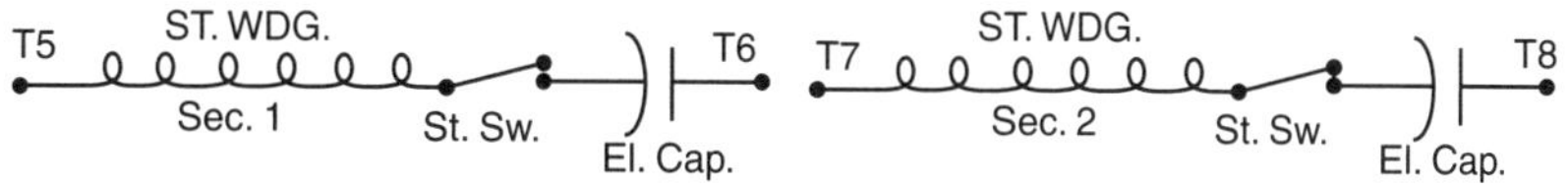

FIGURE 4.10   Dual-voltage start-winding number sequence.

Disconnect the leads T5 through T8. Compare the reading between T5 and T6 to the reading between T7 and T8. (*Do not include the capacitors.*) It may be necessary to turn the shaft slowly while recording the high and low readings. The results should be identical. If one circuit has lower resistance (with an ohmmeter) or higher amperes (with limited AC), it has shorted coils.

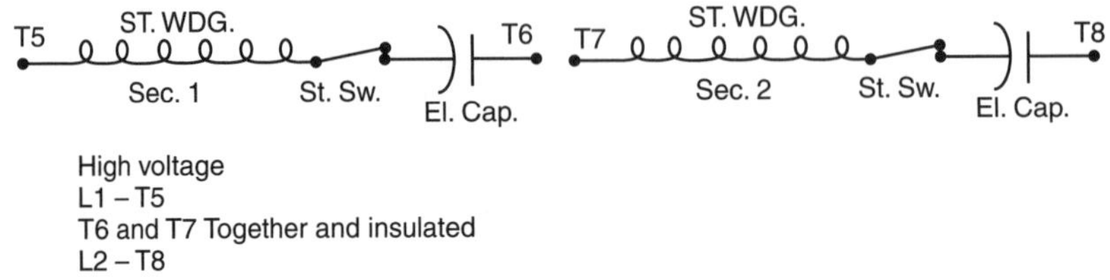

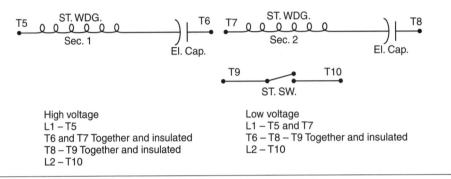

FIGURE 4.11 Dual-voltage start winding (with contacts for each half).

FIGURE 4.12 Dual-voltage start winding with one set of contacts.

Each circuit should also be tested to the frame of the stator for an insulation breakdown to ground. A short to ground requires a complete rewind or replacing the motor.

Start Winding Shorted to Run Winding The start winding and run winding normally aren't internally connected. The only exceptions are motors that have predetermined rotation, such as centrifugal pump motors. If there is continuity between the start and run windings, the motor will have to be rewound or replaced.

Connecting the Dual-Voltage Start Winding Figure 4.13 shows the start-winding lead number system. To connect a dual-voltage start winding, simply use the following phrase: in on the odds and out on the evens. Start-winding leads are always numbered T5 through T8.

Run Winding The run winding normally has fewer problems than the rest of the capacitor-start motor's components. Overheating from overload, high ambient temperature, low voltage, and bearing failure cause the following problems:

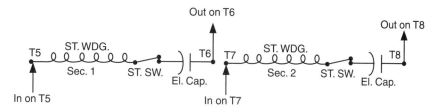

FIGURE 4.13   Connecting a dual-voltage start winding using the in-on-the-odds, out-on-the-evens method.

- Burned coils
- Shorted turns
- Shorted coils (dual voltage)
- Short to the start winding
- Short to the frame
- Open winding (dual voltage)

Burned Coils   Burned coils are always visible. They usually have a burned varnish smell. The motor may even run normally in this condition. It definitely should be replaced or rewound.

Shorted Turns   Shorted turns lower the motor's resistance. (Amperes will increase and cause the windings to overheat.) The stator's magnetic balance becomes upset, making the motor noisy. A ringing or high-pitched sound sometimes indicates shorted turns.

A low power factor can be mistaken for shorted turns. If the no-load amps are 10 to 15 percent higher than nameplate amperes and the motor sounds normal, a low power factor may be the cause. Normally, most single-phase motors have a low power factor. With no load, it's common for fractional horsepower motors to draw higher than their nameplate amperes. They often draw fewer amperes when fully loaded. The power factor of almost all AC induction motors improves when they are fully loaded.

Shorted Coils, Dual-Voltage Run Winding   Many capacitor-start motors have dual-voltage run windings (Fig. 4.14). The dual-voltage run winding has two pairs of lead wires numbered T1 through T4. Half of the run-winding circuit is between T1 and T2; the other half is between T3 and T4.

The two circuits have identical data, so they can be comparison tested. It is highly unlikely that both circuits would have identical problems. Use a microhmmeter or limited AC current for this test.

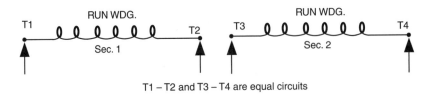

T1 - T2 and T3 - T4 are equal circuits

FIGURE 4.14   Testing both circuits of dual-voltage run winding using the schematic.

Disconnect leads T1 through T4. Compare the readings between T1 and T2 to the readings between T3 and T4. It may be necessary to turn the shaft slowly while recording the high and low readings. The results should be identical. If one circuit has lower resistance (with an ohmmeter) or higher amperes (with limited current), it has shorted coils.

Each circuit should also be tested for a ground to the frame of the stator. Both a short and a ground require that the motor be rewound or replaced.

Figure 4.15 shows the numbering system for run-winding leads. (In 1967, NEMA interchanged the numbers T2 and T3 on single-phase motors.) To connect a dual-voltage run winding, just remember the following phrase: in on the odds and out on the evens. Run windings are always numbered T1 through T4.

Run Winding Shorted to Start Winding   Run and start windings normally aren't internally connected. The only exceptions are motors that have predetermined rotation, such as centrifugal pump motors. If there is continuity between the start and run windings, the motor will have to be rewound or replaced.

Run Winding Shorted to the Frame   Disconnect the power. With an ohmmeter, test from all run-winding circuits to a clean spot on the stator frame. If the reading indicates a solid ground, the motor must be rewound or replaced. If the reading is between 2 and 50 megohms, clean and dry the motor. If this doesn't restore the reading to infinity, the motor must

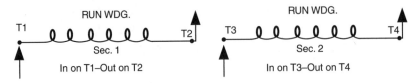

FIGURE 4.15   Connecting a dual-voltage run winding, using the in-on-the-odds, out-on-the-evens methods.

be rewound or replaced. The bottom coils are subject to water-related low ohmmeter readings.

Open Run Winding, Dual-Voltage, Low-Voltage Connection   If one circuit of a dual-voltage run winding is open, the motor will lose slightly more than half its power. The motor will start slower than normal unloaded. Depending on its internal connection method, it will either run smoothly or be slightly noisy.

An ohmmeter or test light can be used to identify the open circuit. If the open is in a lead-to-coil or coil-to-coil connection, it can be soldered. If a coil group is open internally, the motor must be rewound or replaced.

Submerged Motor   If a single-phase motor has been submerged in water, but not energized, it may not need rewinding or replacing. Cleaning and baking its windings might be all that's needed. However, the electrolytic capacitor shouldn't be baked in an oven.

The windings should be tested first with an ohmmeter. (A wet winding should never be subjected to a test voltage that could arc through the wet slot insulation.) The baking temperature should not exceed 200°F. The ohmmeter test should read infinity after baking. The windings should be given a coat of air-drying varnish after they have been cleaned, dried, and tested.

When water soaks the slot insulation, a battery action can sometimes be detected. As long as the slot insulation is wet, a small voltage can be read (with a millivoltmeter) between the winding and the frame. A zero reading would indicate that the motor has been baked long enough. If the ohmmeter test now shows infinity, other test instruments such as a megohmmeter or surge tester can be used.

If a motor has been submerged, it should be disassembled, cleaned, and dried as soon as possible. If it has sleeve bearings, the shaft's bearing surface will soon become rust pitted. The sleeve bearing motor should have the oil wick material and oil replaced. (Ball bearings should be replaced.)

The capacitor and start switch should be cleaned and wiped dry, but if there was prolonged submersion, the capacitor should be replaced. (Contaminated water causes the aluminum components inside the capacitor to corrode.) Oil-filled capacitors can be cleaned and wiped dry because they are sealed.

Bearings   A worn sleeve bearing will make sharp clattering or rumbling sounds, especially when the motor starts. Disconnect the belt and move the shaft back and forth in line with the load. There should be no movement. Small motors are usually replaced because replacing sleeve bearings isn't cost effective.

Ball bearings are very noisy when faulty. They can be replaced, but the repair cost must be compared with the cost of a new motor. The age and condition of the motor is part of this decision.

The stationary switch should also be replaced at this time.

Rotor   Cast aluminum rotors are used in most single-phase motors. If they have open rotor bars or end rings, the motors are replaced because it isn't cost effective to repair them.

Loss of torque is a sign of open rotor bars or end rings. The motor won't draw many amperes with no load and will start a load more slowly than normal. Loaded RPM will be lower than the nameplate value. The start and run windings won't show any problems.

Two-Value Capacitor Motor (Capacitor-Start, Capacitor-Run Motor)

The two-value capacitor motor has the same components as the capacitor-start motor (Fig. 4.16). An oil-filled capacitor is added for power factor improvement. The function of the two-value capacitor motor is discussed in Chapter 3 under "Oil-Filled Capacitor Connection in a Two-Value Capacitor Single-Phase Motor."

Troubleshooting methods for the two-value capacitor motor are the same as those for the capacitor-start motor. They are discussed earlier in the chapter under "The Capacitor-Start Motor."

Following are additional problems and troubleshooting methods for the oil-filled capacitor.

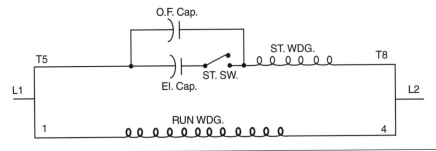

FIGURE 4.16   Schematic of the two-value capacitor motor.

Troubleshooting the Oil-Filled Capacitor

The two major problems for the oil-filled capacitor are open and shorted.

If the oil-filled capacitor becomes open, the full magnetizing amperes of the run winding are carried by the motor's circuit (the circuit's voltage drop will increase). The loss of the oil-filled capacitor won't affect the motor's performance, but it will make more magnetic noise.

If the oil-filled capacitor becomes shorted, the start switch and the electrolytic capacitor(s) are bypassed. The start winding is now connected directly across the line. If the motor is restarted, it draws high current and has very little starting torque. The start winding will be destroyed in a few seconds. Disconnect the shorted oil-filled capacitor, and the motor will operate satisfactorily without it.

Electrolytic Capacitor Connection Rules and Formulas

Capacitors are connected parallel, series, and in combinations of both. The reason for these connections was discussed in Chapter 3 under "Parallel Electrolytic Capacitor Connection" and "Series Electrolytic Capacitor Connection."

Electrolytic capacitor connection rules are as follows:

- Capacitors connected in parallel don't need the same microfarads (mfd) rating.
- Capacitors connected in series must have the same mfd rating.
- There are never more than two capacitors connected in series with each other. Two capacitors in series are normally a high-voltage connection.

The following electrolytic capacitor formulas determine total capacitance value for series and parallel connections and combinations of them:

- **Parallel**   mfd ratings of each capacitor added together = total mfd
- **Series**   mfd rating of one capacitor divided by the number of capacitors in series = total mfd
- **Capacitors of unequal mfd rating in series**   product/sum (example below):

    cap #1 mfd × cap #2 mfd ÷ cap # 1 mfd + cap # 2 mfd = total mfd

In Fig. 4.17 (a) and (b) mfd values are added together for total mfd. In part (c) the mfd value is found by dividing the mfd value of 1 by the total number of them (2).

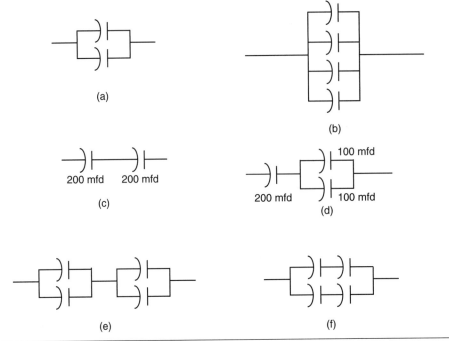

FIGURE 4.17 Six combinations of electrolytic capacitor connections found in capacitor-start motors.

Connection (d) adds the parallel connected capacitor's mfd values and then divides the result by 2.

Connection (e) shows two capacitors in parallel that are in series with two capacitors in parallel. First find each parallel circuit's value and then divide the result by 2 (the number of circuits). (Each parallel circuit must have the same mfd value.)

Connection (f) shows two capacitors in series and in parallel with two capacitors in series. (Capacitors in series must have the same mfd rating.) Divide the value of one capacitor by 2 (the number of them in series) to find the mfd of each circuit. Add the value of each circuit to get the total mfd.

Troubleshooting Electrolytic Capacitors

Figure 4.18 shows how mfd values change when there is a shorted or open capacitor in the connection. The multiplier 11 is explained in Chapter 3 under "Capacitor Test Formula." The number 11 applies because this is a 240-volt application. *The capacitors are all rated 200 mfd and 120 volts.* Using the logic of the test formula, the amperes available to the start winding can be calculated. The results explain why a motor reacts as it does when there are shorted or open capacitors.

The first problem (Fig. 4.18) has two capacitors in parallel and in series with two capacitors in parallel. The second problem (Fig. 4.19) has two capacitors in series, connected in parallel with two capacitors in series. Both connections are used in capacitor-start motors rated 5 hp or more.

In Fig. 4.18, the sum of capacitors #1 and #3 (400 mfd) is the same as the sum of capacitors #2 and #4 (400 mfd): 400 ÷ 2 = 200 mfd. The power is 60 Hz at 240 volts, making the multiplier 11. 200 mfd ÷ 11 = 18 amperes available to the start winding when there are no faulty capacitors.

In Fig. 4.18 (b), capacitor #1 is open. Capacitor #3 (200 mfd) is in series with capacitors #2 and #4 (400 mfd). [200 mfd × 400 mfd] ÷ [200 mfd + 400 mfd] = 133 mfd, and 133 mfd ÷ 11 = 12 amps available to the start winding.

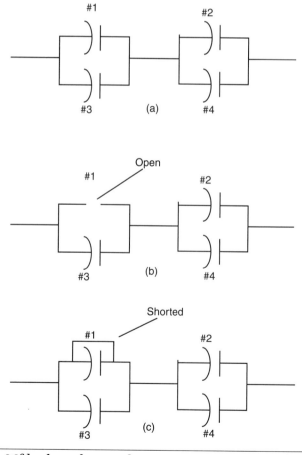

FIGURE 4.18   Mfd values change when a capacitor is open or shorted.

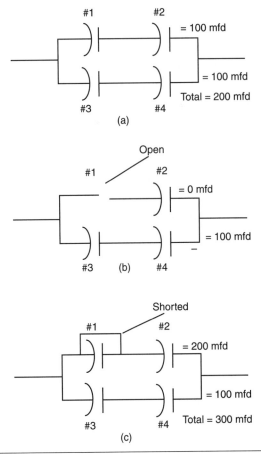

FIGURE 4.19 Mfd values change when one capacitor is open or shorted.

(Starting efficiency is decreased because of the loss of mfd.) The time will decrease between the peak magnetism (phase angle) of the start and run windings. The loss of one-third of the amperes available to the start winding reduces the motor's starting power. The motor won't start a maximum load.

In Fig. 4.18 (c), capacitor #1 is shorted. Capacitor #2 is now in parallel with capacitors #3 and #4. The total capacitance is now 300 mfd.

The phase angle is well off the ideal 90°, lowering the starting efficiency. Start amperes are higher than normal. The motor may start but wouldn't be able to start a load. Extremely high amperes indicates a shorted winding, when actually the problem is only a shorted capacitor.

Motor performance problems that are shown in Fig. 4.19 are similar to problems that were found in Fig. 4.18. This connection is two capacitors in series in parallel with two capacitors in series.

Two-Value Capacitor Motor Connections

Two-value capacitor motors (with electrolytic and oil-filled capacitors) use the connections shown in Fig. 4.20. Oil-filled capacitors are always connected in parallel with the electrolytic capacitor and the start switch contacts. If more than one oil-filled capacitor is used, they are always connected in parallel with each other. The voltage rating of the oil-filled capacitor will be two or more times the applied voltage value. Both types of capacitors are found mounted in the motor, on the motor, or remote from the motor.

Power Factor Correcting Winding   Figure 4.21 shows a schematic of a special winding in series with an oil-filled capacitor. This winding is wound

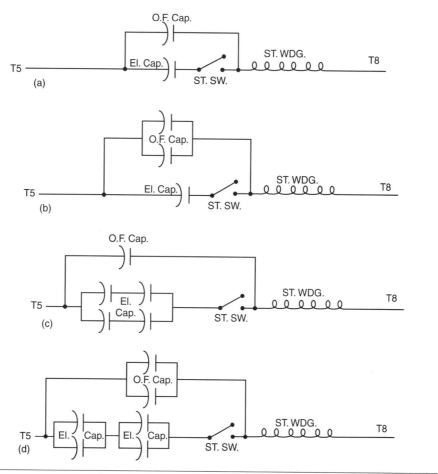

FIGURE 4.20   Several connection combinations found in the two-value capacitor motor's start winding.

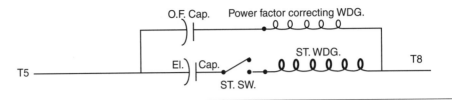

FIGURE 4.21   A separate winding designed for correcting the motor's power factor.

on top of the start winding and connected so their polarities are the same. Its purpose is to correct the power factor of the motor.

High-Voltage Problems of the Two-Value Capacitor Motor   If the two-value capacitor motor is connected to a coasting load (such as a radial arm saw), voltage as high as 1500 volts is generated in the motor's circuitry. (A magnetic interaction between the rotor cage, the closed circuit of the start and run windings, and the oil-filled capacitor creates the high voltage.) The power value of the high voltage is very low, but it can destroy or damage electronic components or anything connected in the motor's circuitry.

Capacitor-Start Motor Connections

Figure 4.22 (a through e) covers most capacitor-start motor connections. There is no standard order of sequence for the components. Regardless of the sequence, the start switch contacts must shut off the electrolytic capacitor(s) and the start winding. Figures 4.22 (a) and (b) can be used on high-voltage, low-voltage, or dual-voltage applications. The capacitor's voltage rating must be the same as the applied voltage when the application is a single voltage. Dual-voltage applications require lower-voltage capacitors. The run winding is used as an autotransformer to keep the voltage applied to the start winding low. Figure 4.23 shows the voltage value when the dual-voltage run winding is connected for high voltage. When the start winding is connected across one-half of the run winding, it will receive only half of the applied voltage.

The series connections in Figs. 4.22 (c) through (e) are all high-voltage applications. The capacitors are all rated low voltage. Figures 4.22 (d) and (e) are connections used on larger-horsepower motors.

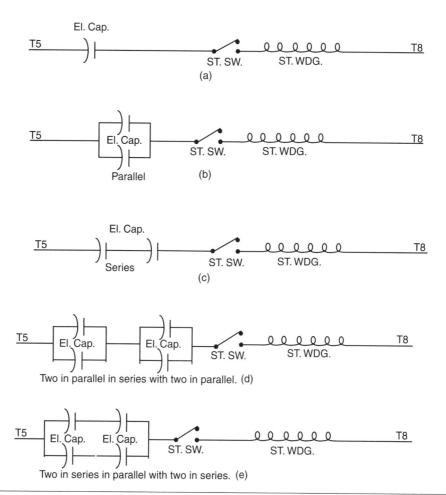

FIGURE 4.22   Start winding connections found in capacitor-start motors. (a) One voltage: This start winding connection can be used on low or high voltage, but not both. The voltage rating of the capacitor determines the voltage applied. (b) One voltage: This start winding connection can be used on low or high voltage, but not both. The voltage rating of the capacitor determines the voltage applied. (c) High voltage only: Capacitors will be rated low voltage only. (d) High voltage only: Capacitors will be rated low voltage only. (e) High voltage only: Capacitors will be rated low voltage only.

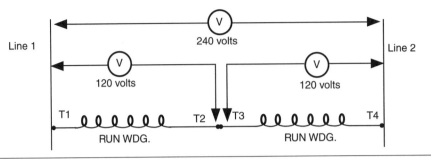

FIGURE 4.23 Schematic showing the voltage across the run-winding coils and leads when the motor is connected for high voltage.

Thermal Protective Devices

Thermal protective devices are used to protect a motor from burnout caused by overload and failure to start. There are two types, manual resetting and self-resetting. Both types have a bimetal component that bends in a predetermined direction when overheated from excessive current (Fig. 4.24).

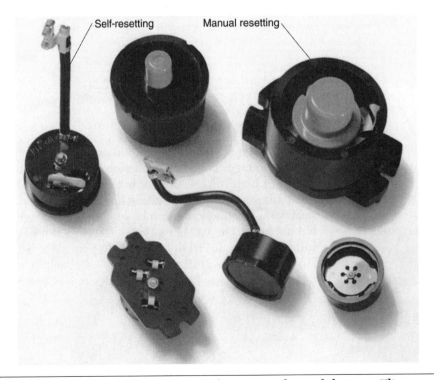

FIGURE 4.24 Self-resetting and manual resetting thermal devices. *Klixon.*

Manual resetting thermal devices must cool before they can be reset. (Larger thermal devices may take 10 to 15 minutes to cool sufficiently.) There is a definite snap when they reset. They should be in a convenient location. If they are mounted in a motor's connection box, they can be relocated.

A self-resetting thermal device will reset itself and start the motor when it cools. Machinery that can cause injury shouldn't have motors with self-resetting protective devices.

Components of the Bimetal Thermal Protective Devices (for Single-Phase Motors)

This type of protective device has a disc and two sets of contacts. Two of the contacts are mounted on the movable part of the disc. The two mating contacts are mounted on the stationary part of the device.

Figure 4.25 shows the circuitry of a thermal device. The terminal marked P1 (on the thermal device) is connected to one of the stationary contacts. A

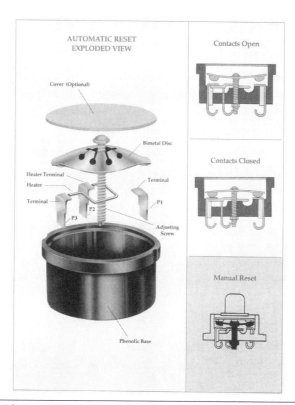

FIGURE 4.25   The internal connection of a thermal device. *Klixon.*

lead in the motor's connection box (labeled P1) is connected to this terminal. If the motor has a connection board, P1 is soldered to a bolt, marked P1. P1 (as a lead or bolt) is always connected to line 1. All of the circuitry of the thermal device and the motor are in series with P1.

The P2 terminal is connected to the stationary contact across the disc from P1. If the motor has a connection box, P2 is a brown wire and may or may not be marked P2. If the motor has a terminal board with connecting bolts, there's also a bolt labeled P2.

Current through P2 has no effect on the function of the thermal device. (It disconnects when the device trips off.) If the lead P2 isn't used in the connection, it must be insulated.

A small heat element (located near the disc) is connected between P2 and P3. When there's an overload, higher than normal current flows through the heat element and P3. The heat element gets hot, making the disc bend, disconnecting all motor circuits. (P3 is always connected to a winding that is affected by an overload.)

Connections and Operation of the Thermal Protective Device

Thermal devices are used on all types of motors. Single- and dual-voltage capacitor-start motor connections with thermal devices are discussed here.

Single-Voltage Capacitor-Start Motor

Figure 4.26 shows a thermal device connected to a single-voltage capacitor-start motor. Line 1 is connected to P1. P3 is connected directly to the run winding. A lead marked T1 is connected to P3, making it accessible to combine start-winding leads T5 or T8 with T1. (To reverse the motor, interchange T5 and T8.) T4 and T5 or T8 are connected to L2.

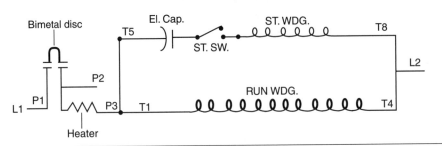

FIGURE 4.26 Thermal device for single-voltage motors.

P1 (line 1) is disconnected from all circuits when the thermal device functions because of an overload.

If the thermal device becomes defective, T1 (with T5 or T8) can be connected directly to line 1. This will bypass the defective thermal device. (The heat element would melt if T1 were inadvertently connected to line 2 when P1 is connected to line 1.)

The start winding is sometimes connected to P2 so that its amperes don't affect the heat element. It could be destroyed at this location if the heat element becomes open.

Dual-Voltage Motor, Connected Low Voltage

The dual-voltage motor's run winding has two circuits. Each circuit contains exactly half of the run winding. This makes it possible to connect the run winding high voltage (series) or low voltage (parallel).

Figure 4.27 shows the voltage value across each pole when the winding is connected high voltage (series). Figure 4.28 shows the same voltage value across each pole when the winding is connected low voltage (parallel). When

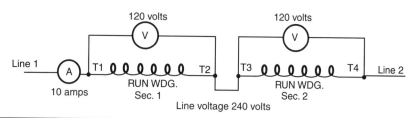

FIGURE 4.27   When connected for high voltage, this motor draws 10 amperes. The voltage read from the T2 and T3 connection to either line is half of the line voltage.

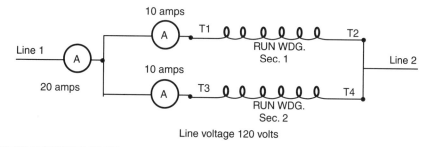

FIGURE 4.28   Each parallel run-winding circuit draws 10 amperes. The voltage drop across each coil is the same as with the high-voltage connection.

the voltage is the same across each pole, the amperes through each pole are the same (for both connections).

Full-load amperes for this motor are 20 amps (low voltage) and 10 amps (high voltage). When connected low voltage (parallel), each circuit draws 10 amps, for a total of 20 amps. When connected high voltage, the run circuit draws 10 amps.

The heat element of the thermal device is connected in series with one circuit of the run winding. Whether connected high or low voltage, the full-load amperes through the heat element will be 10 amps. The trip value for this motor's protective device is based on a 10-amp load.

Figure 4.29 shows a dual-voltage capacitor-start motor connected low voltage. Line 2 is connected to T2, T4, and T5 (or T8). (To reverse this motor, interchange T5 and T8.) Line 1 is connected to P1. T1 (from the run winding) and an external lead labeled T1 are internally connected to terminal P3 of the thermal device.

External lead T1 is sometimes connected to T5 (or T8) and insulated. Some motor nameplates specify that T5 (or T8) be connected to P2. (The start-winding current won't go through the heat element when connected to P2.)

Run-winding lead T3 is connected to P2 and insulated. The current of T3 (10 amps) will bypass the heat element when connected to P2. If the thermal device opens because of an overload, T1 and T5 (or T8) are disconnected from line 1. At the same time, P2 disconnects T3 from line 1. T2, T4, and T5 (or T8) remain connected to line 2.

Dual-Voltage Motor, Connected High Voltage

Figure 4.30 is the same motor as described above, but connected for high voltage. At full load, the run winding draws 10 amperes. Line 1 is connected

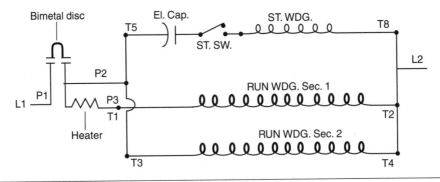

FIGURE 4.29 A dual-voltage motor—and its thermal protection device—connected for low voltage.

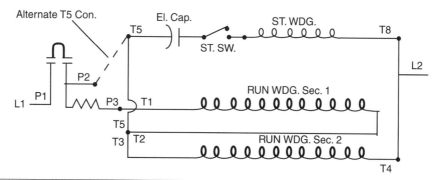

FIGURE 4.30   The same motor and thermal device connected for high voltage.

to P1. P2 isn't used and is insulated, and T2, T3, and T5 (or T8) are joined and insulated. T4 and T5 (or T8) are connected to line 2. (T5 or T8 can also be connected to P2.) This motor can be reversed by interchanging T5 and T8.

Whether this motor is connected high or low voltage, each run winding pole must have 10 amps flowing through it to produce full power.

Limitations of Thermal Protective Devices

Protection that is provided by the thermal device can be reduced by temperature variations. Ambient temperature affects the trip time of all thermal devices. If the ambient temperature is 10°F, the device may not trip soon enough to protect the motor.

Thermal protective devices are mounted either inside the motor or in its connection box.

When the device is mounted inside (Fig. 4.31), the motor's internal temperature is factored into the thermal device's trip temperature. The trip temperature is based on the motor having reached full running temperature.

If the motor is overheated, it may trip off each time the motor tries to start. The cycle will continue until the motor is disconnected long enough for both to cool. If the motor is extremely cold, the thermal device may not trip before the windings are damaged.

Frequent starts can cause excessive tripping with an internally mounted thermal device. High ambient temperature may trip off a lightly loaded motor when it is restarted.

If the thermal device is mounted in the connection box, the motor's internal temperature won't affect it.

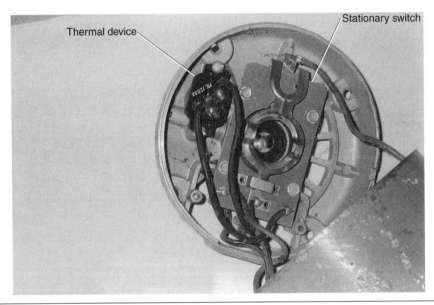

FIGURE 4.31 Thermal device mounted in the motor's end frame. *Marathon Electric.*

Alternative Start-Winding Switches

The stationary switch and centrifugal device assembly can be replaced with numerous types of start-winding controls, which include:

- Potential relay start switch
- Electronic start switch
- Time-delay electronic start switch
- Heat-activated start switch
- Current relay start switch

Potential Relay Start Switch

The potential relay start switch is shown in Fig. 4.32. A motor's original stationary switch contacts and centrifugal device can be replaced with a potential relay. They are available for 115- and 230-volt applications. The same switch can control two-, four-, and six-pole capacitor-start motors.

Potential Relay Components

The potential relay start switch illustration in Fig. 4.33 consists of a coil and a set of horsepower-rated contacts. The coil must match the voltage applied

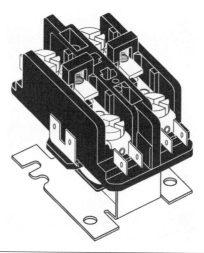

FIGURE 4.32 A potential relay used for the start-winding control. *White-Rodgers.*

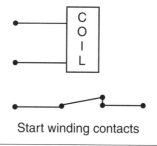

Start winding contacts

FIGURE 4.33 The potential relay's four connections.

to the motor (115 or 230 volts). A dual-voltage motor requires a coil rated for a 115-volt application.

The relay contacts must be rated as high as or higher than the motor's horsepower. Some large capacitor-start motors have a two-circuit start winding and use a separate relay for each circuit.

Operation of the Potential Relay

Figure 4.34 is a schematic of the motor and its potential relay. The power factor of the start winding is low when the motor is starting. As the rotor accelerates, the voltage/current angle (timing), or power factor, of the start winding improves. As the power factor improves, the voltage transformed from the rotor poles to the start winding increases. At 75 to 80 percent of synchronous speed, the voltage across the start winding is high enough to

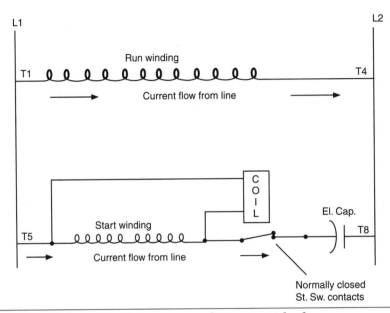

FIGURE 4.34 Schematic of the motor and its potential relay.

make the relay coil function. The start winding is now disconnected from the line. Power continues to be transformed into the start winding from the rotor, keeping the contacts open until the motor is shut off. (A voltage is transformed into the start-winding coils from the poles of the rotor.) Figure 4.35 shows the alignment of rotor poles.

The voltage value (at 75 to 80 percent of synchronous speed) varies slightly from motor to motor. Spring tension (which holds the contacts closed) can be adjusted to fit a motor's individual requirement. Increasing or decreasing the spring tension is done by bending the spring support.

If the motor's capacitor is connected in the relay coil/start-winding circuit, the power factor is immediately raised, and the relay coil functions too soon (Fig. 4.36). When the contacts open, the rotor speed is so slow that the run winding can't pull the rotor up to speed.

If the motor is extremely overloaded, or if there is a short power interruption, the potential relay reconnects the start winding and brings the rotor back up to disconnect speed. (Overload protection should disconnect the motor before its windings are damaged by an overload.)

Coasting Load Problem with Potential Relay

A coasting (or flywheel type) load will generate power in the motor's circuitry. As the motor coasts, the magnetized rotor, the stator windings, and

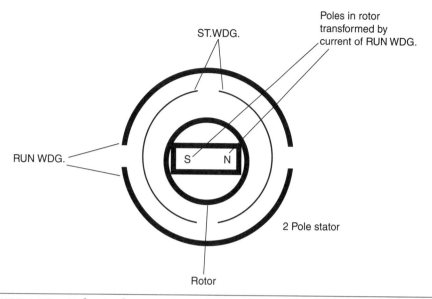

FIGURE 4.35 Poles in the rotor iron align with the start winding.

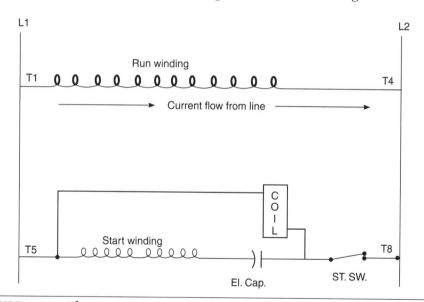

FIGURE 4.36 This connection won't allow the motor to start a load.

the capacitor(s) change the motor into a static generator. Generated power (within the windings) will cause the potential relay to open and close the switch contacts several times before the motor stops.

When the motor is shut off, the relay coil allows the start contacts to close. With the contacts closed, the run winding provides a closed circuit, which allows the amperes to flow. Amperes (flowing in the run winding) create poles in the rotor cage that magnetize the rotor iron. Current flow through the rotor cage transforms a voltage into the start winding. The transformed voltage (helped by the discharging capacitor current) in the start winding is high enough to open the contacts again. This cycling will continue until the speed no longer produces enough voltage to make the relay coil function.

There is excessive wear on the relay's contact assembly when the contacts cycle on and off. Figure 4.37 shows a connection that will eliminate this problem. When the start winding is isolated by this connection, the capacitor doesn't discharge into the closed circuit of the stator windings and there is no contact cycling. A 15,000-ohm, 2-watt bleeder resistor should be installed across the capacitor to discharge it, when the motor is idle. The resistor alone will not keep the switch from cycling.

Electronic Start Switch

The electronic start switch is a very popular replacement switch. Many motor manufacturers use this switch on their single-phase motors. Many junked single-phase motors could have been repaired with these switches, for a fraction of their replacement cost.

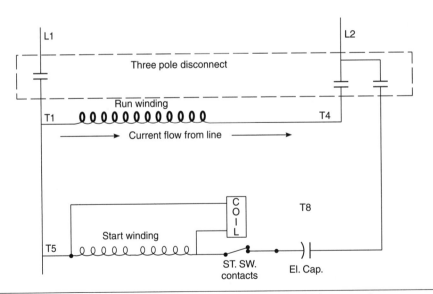

FIGURE 4.37 This three-pole switch connection will eliminate the "open and close" cycling of the contacts.

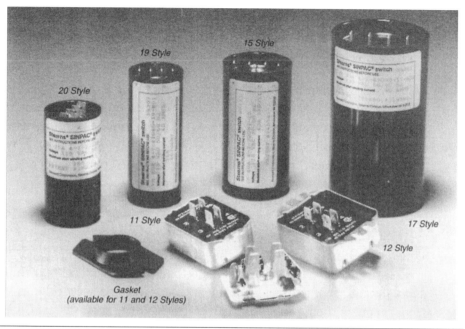

FIGURE 4.38   Electronic switches. *Sinpac.*

Electronic switches operate on the same principle as the potential relay. Figure 4.38 shows some of the switches and their mounting assemblies.

The circuitry of the solid state electronic start switches is completely potted and sealed, making them ideal for dirty locations. Because they have no moving parts, they aren't affected by vibration or extra contact closure during motor coastdown.

This switch has voltage-sensing circuitry that switches off the start winding at a preset voltage. If the motor is overloaded or if there is a power interruption, the switch will reconnect the start winding at about 50 percent of synchronous speed.

Electronic switches are made for split-phase, capacitor-start, and two-value capacitor motors. The same switch will control two-, four-, and six-pole motors. The only two requirements are that the switches have the right voltage rating and a large enough ampere rating.

The electronic switch will control any size motor that draws less than the switch's ampere rating. (It's very important that the ampere rating of the switch is as high as or higher than the motor's nameplate ampere value.)

The cutout voltage values for 115-volt switches are 130, 147, and 165 volts. The cutout voltage values for 230-volt switches are 260 and 310 volts.

To find the required voltage, run the motor with no load on its rated voltage. Isolate the start winding from the line and record the voltage across it (T5 and T8). Select a switch that has a voltage value rating closest to 80 percent of the voltage found across T5 and T8. (Be sure the ampere rating of this switch is equal to or higher than the motor's full load amperes.)

If the motor has a coasting-type load, (for example, a radial arm saw), a special resistor is used to prevent high voltage from destroying the electronic components during coastdown.

These switches replace the switch contacts and centrifugal devices for all split-phase, capacitor-start, and two-value capacitor motors.

Time-Delay Electronic Start Switch

Time-delay electronic start switches shut off the start winding at a predetermined time. They are available in $1/3$ second, $1/2$ second, and 1 second delay time. If the motor stalls or doesn't get up to speed by the time the contacts open, the motor must be restarted to reconnect the start winding. (If the thermal device doesn't trip, the run winding will be destroyed.)

It is possible to plug (reverse while running) the motor with this switch. (A motor with a centrifugal device has to slow enough for the contacts to close before it can be reversed.)

The same switch will operate capacitor-start, split-phase, and two-value capacitor-start motors of two to eight poles. Horsepower ratings are from the smallest motor to 5 hp. The voltage ratings of these switches are 115 and 230 volts. Ampere ratings are 30 amperes and below.

These switches will replace a motor's stationary contacts and centrifugal device.

Heat-Activated Start Switch

Many sealed refrigeration compressors have heat-activated start switch controls (Fig. 4.39). When the motor starts, the start-winding current flows through a bimetallic strip. The bimetallic strip quickly heats, bends, and opens the start-winding contacts.

A small heat element next to the bimetallic strip is connected parallel to the contacts and in series with the start winding. It bypasses the start switch contacts. After the switch contacts open, a small current through the start winding heats the element and keeps the bimetallic strip hot. As the motor runs, the bimetallic strip stays hot enough to keep the switch contacts open.

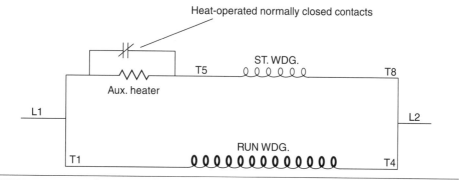

FIGURE 4.39 Schematic of a heat-operated start-winding switch.

The motor won't start if it's restarted before the bimetallic strip cools and reconnects the start-winding contacts. A thermal protective device should trip and shut off the power. The thermal protective device is usually self-resetting, and by the time it cools off enough to restart the motor, the heat-activated start switch has reset.

Current Relay Start Switch

The current relay start switch is used on single-phase motor applications where a stationary switch and centrifugal device aren't practical. (A submersible deep well water pump is a good example.) This application requires that the motor and pump be pulled up from the bottom of the well to be serviced if the start winding controls are mounted in the motor.

A current relay start method makes it possible to locate the start switch contacts and capacitor where they are accessible. (Capacitors and switch contacts need servicing more frequently than the rest of the motor.) Another advantage the current relay has is that only three cables are needed to power the pump motor and control the start winding.

Current Relay Start Switch Connections

The current relay switch in Fig. 4.40 is connected to the three motor terminals marked C (common), S (start), and R (run). The common lead (C) connects to one side of the start winding and one side of the run winding. The other side of the run winding (R) is in series with the relay coil and line 2. The start winding (S) is in series with the switch contacts, the capacitor, and line 2.

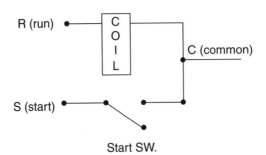

FIGURE 4.40   Current-operated start switch.

Function of the Current Relay Start Switch

The run-winding current continuously flows through the current relay's coil. When the motor starts, high run-winding current pulls the (normally open) start contacts closed (Fig. 4.41). The start-winding circuit is now connected to line 2, and the motor starts. As the rotor accelerates, ampere flow becomes lower in the run-winding and relay coil. At 75 to 80 percent of synchronous speed, the relay coil can no longer hold the contacts closed. The contacts drop open and the start winding is disconnected. The run winding will now pull the rotor up to full speed (Fig. 4.42).

The relay coil's ampere rating must fit a motor's run-winding ampere demand. A motor that's too small won't draw enough amperes to close the switch contacts. A motor that's too large will draw so many amperes that the contacts won't drop open.

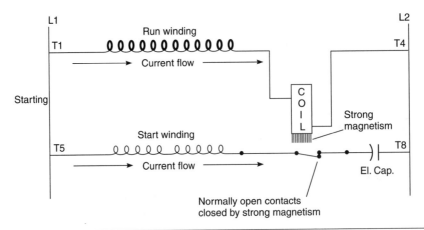

FIGURE 4.41   Schematic of the current relay as the motor starts. The current of the run winding (through the coil) pulls the contacts closed.

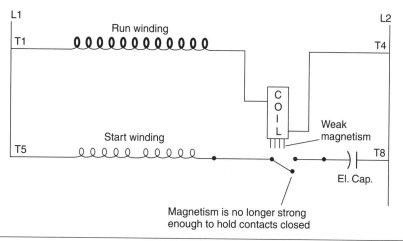

FIGURE 4.42   Current relay contacts in the run position.

Most current relay switch contacts fall open by gravity. For this reason, the relay must not be mounted at an angle. Even a slight tilt will prevent the contacts from falling open. All-angle current relays have spring-loaded contacts. A relay stamped "this side up" isn't all-angle.

A pulsing-type load may cause the contacts to close every time the load peaks. This overheats the start winding and the capacitor and causes the switch contacts to prematurely fail.

Chapter 4 Review

1. The start winding of a shaded-pole motor is a single copper band (page 155). T____ F____

2. The speed of the shaded-pole motor is determined by the load (page 156). T____ F____

3. The split-phase motor has both start and run windings across the line when it starts (page 158). T____ F____

4. Both start and run windings can be energized continuously (page 158). T____ F____

5. The split-phase motor has less starting torque than the capacitor-run motor (page 159). T____ F____

6. Most split-phase motors are replaced—rather than repaired—if there is a winding failure (page 158). T____ F____

7. The capacitor-run motor operates continuously with both start and run windings in the circuit (page 159). T____ F____

8. How many moving parts are in the capacitor-run motor (page 159)?

9. The multispeed capacitor-run motor's RPM is governed by the speed selection, regardless of the load (pages 159–160). T____ F____

10. The capacitor-run motor is usually replaced if there is a winding failure (page 161). T____ F____

11. The capacitor-start motor has high starting torque because of its electrolytic capacitor (page 161). T____ F____

12. The start switch is usually the first component to fail in the capacitor-start motor (page 162). T____ F____

13. The main cause(s) of dead spots are (page 163)
    a. faulty contacts.
    b. a worn rotor device.
    c. worn thrust washers.
    d. all of the above.

14 If thrust washers keep the rotor out of alignment with the stator (off magnetic center) the motor's amperes decrease (page 163). T____ F____

15. It's more economical to replace the original start switch mechanism of the capacitor-start motor with an electronic start switch, because it has no moving parts to adjust (page 164). T____ F____

16. DC current will slow a coasting squirrel cage rotor, but not lock it (page 164). T____ F____

17. A continuous high ohmmeter reading means a capacitor is (page 164)
    a. open.
    b. shorted.
    c. has no problem.

18. Voltage stored in a capacitor can damage an ohmmeter (page 164). T____ F____

19. A motor with an open capacitor won't start (page 165). T____ F____

20. If an electrolytic capacitor is open or shorted, the motor won't start (page 165). T____ F____

21. Overheating will weaken an electrolytic capacitor (page 165). T____ F____

22. A start winding will overheat if not shut off within 3 seconds (page 166). T____ F____

23. A capacitor-start motor will be insulated better than when it was new if it is rewound by a reputable electric motor service center (page 166). T___ F___

24. A dual-voltage start winding has three equal circuits (page 166). T___ F___

25. The number sequence for connecting a dual-voltage single-phase motor can be remembered using the phrase "in on the _____s, out on the _____s." (page 168)

26. A single-phase motor normally has a low power factor (page 169). T___ F___

27. A low power factor can be mistaken for shorted turns (page 169). T___ F___

28. The data between T1 and T2 is the same as between T5 and T8 (page 170). T___ F___

29. The comparison test is the best way to check for shorted turns in a dual-voltage run winding (page 170). T___ F___

30. The term "in on the odds and out on the evens" applies to both the start and run windings (page 170). T___ F___

31. A dual-voltage capacitor-start motor won't start or run with an open in one of its run circuits (page 171). T___ F___

32. A motor that has been submerged in water must be replaced (page 171). T___ F___

33. What is the purpose of the oil-filled capacitor in a two-value capacitor-start, capacitor-run motor (page 173)?

34. The start winding will be destroyed if the oil-filled capacitor is shorted (page 173). T___ F___

35. If the oil-filled capacitor is removed, the two-value capacitor-start, capacitor-run motor won't start (page 173). T___ F___

36. What are the three rules for electrolytic capacitor connections (page 173)?

37. How many electrolytic capacitors can be connected in series (page 173)?

38. Electrolytic capacitors connected in series are always used on high voltage, and the single capacitor is always used on low voltage (page 173). T___ F___

39. When one capacitor is open in a multiple-capacitor system—two in parallel in series with two in parallel—what are the two factors that decrease the motor's starting power (pages 175–176)?

40. When one capacitor is shorted in a multiple-capacitor system—two in parallel in series with two in parallel—the total capacitance is increased (page 176). T____ F____

41. In the term "phase angle"—phase meaning winding, angle meaning the time current flows—a comparison is being made between the time current peaks in one winding (phase) and the time (angle) the current peaks in another winding (page 176). T____ F____

42. When the degrees of the phase angle and the degrees of separation between two windings (phases) match, the motor's torque is at its highest (page 176). T____ F____

43. The voltage rating of an oil-filled capacitor should be the same as the applied voltage (page 177). T____ F____

44. Voltage generated when a two-value capacitor motor coasts can become destructively high (page 178). T____ F____

45. Dual-voltage motors always have capacitors rated for high voltage (page 178). T____ F____

46. Capacitors connected in series are always rated for high voltage (page 178). T____ F____

47. Fractional horsepower single-phase motors with worn bearings are usually replaced rather than repaired (page 181). T____ F____

48. Thermal protective devices (that are self-resetting) shouldn't be used on machinery that can cause injury (page 181). T____ F____

49. Overload current flowing through P2 causes a thermal protective device to trip (page 182). T____ F____

50. P3 is always connected to a winding circuit (that will draw high current) when the motor is overloaded (page 182). T____ F____

51. The thermal protective device will be destroyed if P3 is connected to line 1, when P1 is connected to line 2 (page 183). T____ F____

52. P2 is always connected to windings that should bypass the heat element (page 183). T____ F____

53. When a dual-voltage motor is connected low voltage, only half of the run winding is connected to P3 (page 184). T____ F____

54. A fully loaded dual-voltage motor will have the same number of amperes through its run-winding coils on low voltage as it has on high voltage (pages 184–185). T____ F____

55. High ambient temperature can cause thermal protective devices to (page 185)
    a. not trip.
    b. trip needlessly.
    c. all of the above.

56. The potential relay start switch can replace a motor's stationary switch and rotor device, regardless of its speed (page 186). T___ F___

57. The voltage across the run winding controls the potential relay's function (pages 186–187). T___ F___

58. The motor's horsepower rating determines the size of the potential relay's contacts (page 187). T___ F___

59. Voltage across the start winding will increase to a value above line voltage when the rotor reaches 75 to 80 percent of synchronous speed (pages 187–188). T___ F___

60. A coasting load causes excessive wear on the potential relay's components (pages 188–189). T___ F___

61. A bleeder resistor will keep a potential relay from cycling (pages 188–189). T___ F___

62. An electronic switch—that operates the same as the potential relay—has the same voltage/horsepower restrictions (pages 190–191). T___ F___

63. A coasting load won't affect the electronic switch (page 192). T___ F___

64. A time-delay electronic switch operates regardless of the motor's speed (page 192). T___ F___

65. A heat-activated start switch is compatible with refrigeration compressors, because a compressor will stay off long enough for the switch to reset (pages 192–193). T___ F___

66. A thermal protective device is vital when the heat-activated switch is used (page 193). T___ F___

67. The current relay depends on the amperes of the run winding to function (page 194). T___ F___

68. The current relay can't close the contacts if the motor is overloaded (page 194). T___ F___

69. The current relay's ampere rating has to match that of the motor (page 194). T___ F___

Chapter 5

The Three-Phase Induction Motor

Industry uses more three-phase induction motors than other types of motors because they require less maintenance. The squirrel cage rotor is the motor's only moving part (excluding ball bearings). This motor is smaller physically (per horsepower) than single-phase or DC motors (Fig. 5.1). Sizes range from fractional to thousands of horsepower (Fig. 5.2).

Three-phase power is divided among three lines. This reduces the supply line's wire size.

The three-phase induction motor has no switching or commutation of circuits. This makes it possible to use very high voltage. High voltage allows the motors to be designed in large sizes (as volts go up, amperes go down—watts and horsepower stay the same).

The speed of the three-phase induction motor is governed by the number of poles in the stator and the hertz of the power source (like the single-phase induction motor).

Components of the Three-Phase Motor

Figure 5.3 shows all the three-phase motor's components:

- Rotor and fan
- Stator core
- Shell and frame
- End brackets
- Bearings

FIGURE 5.1 Small three-phase induction motors. *Lincoln Electric Motors.*

FIGURE 5.2 A large three-phase synchronous motor. *Bradley Electric.*

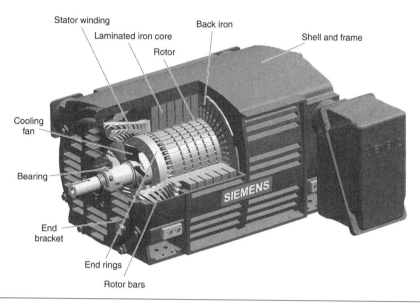

FIGURE 5.3   A three-phase induction motor with all its components shown. *Siemens.*

Rotor and Fan

The squirrel cage rotor in Fig. 5.3 is cut open to show some of its rotor bars. The rotor bar's size, shape, and effect are covered in depth later in this chapter, starting with "Amperes and Rotor Bar Design."

The rotor has a cooling fan attached to it. Fans can be located internally, externally, or both. Air moving over the windings and core iron removes most of the heat from a motor. (The rest escapes through the motor's shell.) Most cast aluminum rotors have fan blades cast in their end rings. Ventilation is the same for either direction of rotation.

The motor's location should not restrict intake or exhaust air movement.

Stator Core

The stator core is made of thin steel sheets called laminations. A laminated core is necessary to reduce (heat-causing) eddy current. The thickness of the laminations is governed by the frequency for which the motor is designed. (Higher frequency develops greater eddy current and requires thinner laminations.)

The steel used in laminations is very permeable, meaning that it changes polarity with minimal internal friction. Magnetism is constantly changing in the laminated core of the three-phase motor.

Slots are cut in the laminations with a laser or punch press. The slotted laminations are stacked to a thickness required for the motor's horsepower rating.

The iron below the slot (back iron) (Fig. 5.3) completes the magnetic path from pole to pole. A two-pole motor will have much thicker back iron than a six-pole motor. The AC motor (unlike the DC motor) doesn't use its outer shell to complete the magnetic path.

Some motors have ventilation slots between the core and the motor's shell. The outside or back iron of the (laminated) core should be as symmetric as possible, because it affects the motor's magnetic path.

Shell and Frame

The motor's shell is made of cast iron or pressed steel. The shell supports the end frames (end bells), which contain the motor's bearings.

Most motors have mounting feet that are part of—or are welded to—the shell. The c-flange design is mounted from the shaft end bell (or end frame).

A cast iron frame is more rigid than a pressed steel frame, making it better for motors of 10 horsepower and larger. Cast iron components, however (such as mounting feet and end bells), can break from rough handling.

Properly designed pressed steel frames are better than cast iron for motors under 5 horsepower. Their lighter weight makes them easier to handle.

Short compact motors are stronger than long motors. The rotor of a short-framed motor has less distance between its bearings, lessening the chance of rotor drag. A short compact motor handles rough duty (such as a chain drive or a direct driven vibrator) very well.

Electric motors have NEMA and European standard frame dimensions. The frame dimensions include shaft size, height, length, and mounting hole locations.

End Brackets

The motor's end brackets are made of cast iron, cast aluminum, or pressed steel. They support the sleeve (or ball) bearings.

The end brackets of large sleeve bearing motors contain an oil reservoir. (An oil ring rides on the shaft and brings oil to it from the reservoir.) Small sleeve bearing motors have oil wicking which is retained around the bearing in the end bracket.

End brackets with ball bearings may or may not have allowances for greasing the ball bearings. This is covered in detail under "Bearing Maintenance" in Chapter 7.

Ventilation slots are part of the end bracket if the motor is classified as open. An open motor can be open, dripproof, or splashproof. Windings can be seen in an open motor. A dripproof motor protects the windings from falling water. Splashproof motors have baffles that shield the windings from above and below, but still allow free movement of air through the motor.

Totally enclosed, fan-cooled frames are dustproof and have no ventilation slots in the end brackets. The shaft is extended through the rear end bracket with a fan mounted on it. (A guard shell covers the fan.)

Explosion-proof motors are built strong enough to contain an explosion of gasoline vapor, without emitting flame or sparks into the surrounding air.

Bearings

Electric motors have ball, roller, or sleeve bearings. The majority of motors have ball bearings. Some large motors and special-duty motors have roller bearings. Bearings are covered in detail under "Bearing Maintenance" in Chapter 7.

Operation of the Three-Phase Motor

The operating characteristics of the three-phase induction motor are directly linked to the three-phase alternator that powers it. The positioning of the stator poles in both is identical. The voltage sine wave (Fig. 5.4) produced by the three-phase alternator is a picture of how the poles are connected in both the motor and the alternator.

The alternator has three single-phase windings, spaced 120° apart in its stator (Fig. 5.5). Figure 5.6 shows the coils that form the first pole of phase A, phase B, and phase C. If a magnet is passed over the coils, the voltage generated in them produces a sine wave like the one shown in Fig. 5.7. The voltage produced is spaced 120° apart (like the coils). At 60 Hz, 120° in time equals 1/280 second. Voltage peaks in each of the three phases 1/280 second apart. Figure 5.8 shows the polarity of six three-phase poles and how they fit the sine wave.

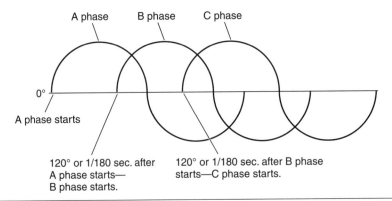

FIGURE 5.4 The three-phase voltage sine wave. The shape of the sine wave makes it easy to compare with the poles of a stator.

FIGURE 5.5 The stator of a three-phase alternator has windings that are identical to a three-phase motor. *Smith Services.*

The Rotating Magnetic Field in a Three-Phase Motor

A three-phase motor has three single-phase windings. They are positioned 120° apart, identical to those of the three-phase alternator previously described. The following is a simplified version of the three-phase rotating magnetic field.

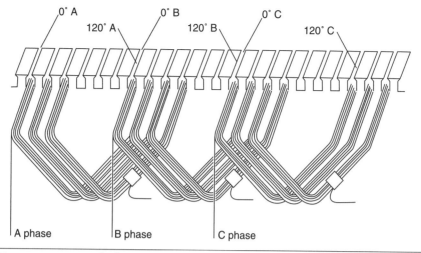

FIGURE 5.6   One pole from each of the three phases. (Poles are separated 120 electrical degrees from each other.)

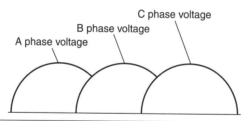

FIGURE 5.7   Voltage sine wave created by the three poles (if a magnet is passed over them).

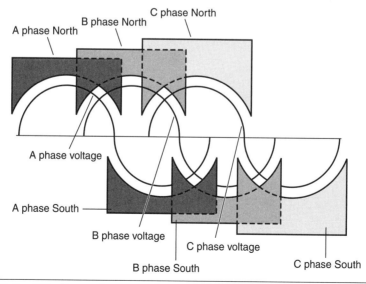

FIGURE 5.8   Windings of a two-pole stator are connected to fit the sine wave.

Figure 5.9a is a two-pole stator with a compass needle centered in it. When DC is applied to phase A, the compass needle will align as shown in Fig. 5.9b. Applying DC to phase B moves the compass needle 120°, aligning it with phase B (Fig. 5.9c). And DC applied to phase C moves the compass needle another 120°, aligning it with phase C (Fig. 5.9d).

Applying DC as shown in Fig. 5.9 creates a rotating magnetic field (like three-phase power). Rotation is produced when three voltages (spaced 120° in time) are applied to three phase windings (spaced 120° from each other). Alternate phases will reach peak magnetism at 1/280-second intervals,

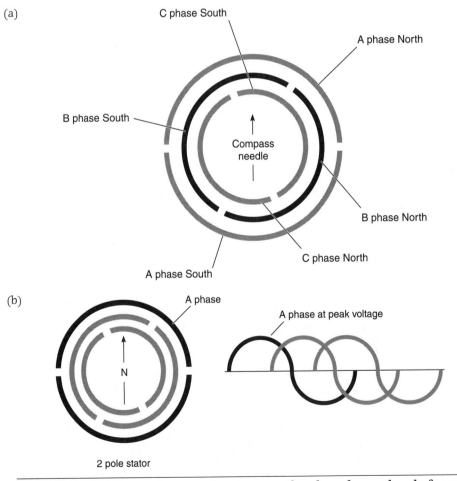

(a)

C phase South

A phase North

B phase South

Compass needle

B phase North

C phase North

A phase South

(b)

A phase

A phase at peak voltage

N

2 pole stator

FIGURE 5.9   (a) Two-pole stator showing a north pole and a south pole for each phase. (b) When DC is applied to the A phase, the compass aligns itself as shown.

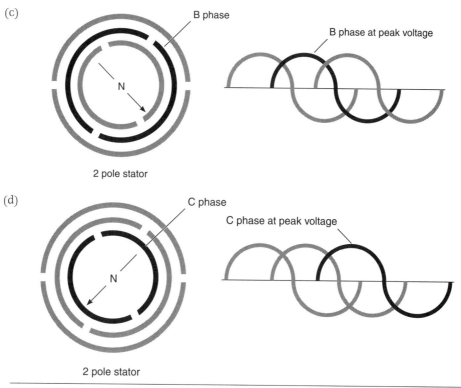

FIGURE 5.9   (c) The compass needle moves 120 degrees when DC is applied to the B phase. (d) When DC is applied to the C phase, the compass needle moves another 120 degrees.

resulting in a rotating magnetic field. The moving magnetic lines of force cut the bars of the rotor's squirrel cage winding, creating poles in the rotor iron.

Single-Phase Condition

A three-phase motor won't start with one of its lines open. When one line is open, single-phase is applied to two of the motor's three windings and one winding is inactive. Two of the three single-phase windings (spaced 120° apart) are energized. Current in the two windings is in unison. Magnetism of both windings peaks at the same time, as though they were one winding. There's no rotating magnetic field. Since the rotor pole magnetism has no stator pole magnetism to react to, there is no torque. A pulsing, high current will flow in the rotor and the stator windings.

If the shaft is spun fast enough, the motor will run as a single-phase motor. The motor's power, however, drops to slightly more than one-half its nameplate value. If a fully loaded motor is running when single phasing occurs, the energized part of its winding overheats and chars. This is explained and illustrated in Chapter 6, Fig. 6.35.

Amperes and Rotor Bar Design

The motor's amperes and torque are affected by various designs of the rotor bars. The type of metal used, the bar shape, and size all affect the motor's amperes and torque (from zero to full speed).

Cast aluminum is the most common metal used in squirrel cage rotors. Aluminum makes it easier and cheaper to mass-produce rotors. However, because of the low melting temperature of aluminum, rotor bars and end rings have to be oversized to carry the high starting current. (This reduces the amount of iron that is needed for the magnetic lines of force.) The overall size of the rotor must be larger when aluminum is used instead of copper or brass.

Rotor bars made of copper (and/or other alloys) conduct high-ampere flow (that would melt aluminum). An example is the small copper rotor bars that are used in high-slip motors. Their small size gives them high resistance, which limits the amperes in the rotor winding.

Ampere flow in all motor windings is very high when a motor starts. The motor's current demand (the instant it starts) is the same as the locked rotor current. The squirrel cage winding (with locked rotor) is similar to a shorted secondary in a transformer. (If both ends of each turn in a transformer's secondary were welded to two separate bars of copper, the resulting ampere flow would resemble that of a locked squirrel cage rotor.) When the rotor is locked, the bar size and the resistance of the bar's metal are all that control the current flow in its winding. (Large rotor bars have greater area for current flow than small rotor bars.) Ampere demand can be many times the motor's nameplate value.

When the motor is at 0 RPM (locked rotor), the slip is 100 percent. (Percent of slip is synchronous speed minus the rotor speed divided by synchronous speed times 100.) At this time, the rotating magnetic field is cutting the squirrel cage winding bars at maximum hertz. As the speed of the rotor increases, slip decreases, and the rotor hertz decrease. The power supply hertz times percent of slip equals rotor hertz. Rotor hertz ranges from 0.3 to 1.5 Hz at full-load speed (depending on the rotor bar design). When the rotor hertz is this low, fewer lines of force cut the rotor bars, so less power is transformed into them. Since the rotor's power comes from the stator, the stator amperes drop to normal (at rated speed).

Phase Angle

Phase (in the term *phase angle*) means a winding. (The squirrel cage is the rotor's winding.) *Angle* means time.

Angle, in this case, is a reference to the time that the current flows in two respective windings and creates magnetism. *Phase angle is the difference (in time) between when one winding and another winding reach peak magnetic power.*

The phase angle has a bearing on a motor's torque. A motor with a good phase angle (when starting) has high starting torque, but not necessarily a high ampere demand. High amperes don't always mean high starting torque.

Small Rotor Bars

When a squirrel cage induction motor is energized, there's an instant of locked rotor condition. At this time, magnetic lines of force of the rotating magnetic field penetrate the rotor's iron only about $3/8$ inch. A voltage is transformed into small rotor bars located within this $3/8$ inch (Fig. 5.10). The ampere flow in the rotor poles (transformed amperes) and the ampere flow in the stator poles (line amperes) will peak in value at nearly the same time. When this happens, the motor has a good phase angle (near $90°$) between stator and rotor windings.

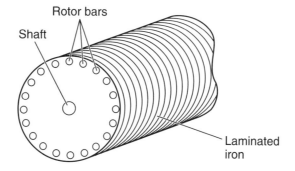

FIGURE 5.10   Small rotor bars, located close to the surface of the rotor, will have ampere flow that nearly matches (in time) the stator amperes.

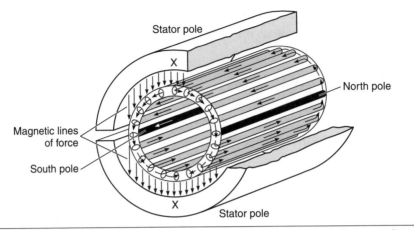

Stator pole

North pole

Magnetic lines of force

South pole

Stator pole

FIGURE 5.11 Poles form in the rotor's iron as current flows in the rotor bars.

Poles form in the rotor iron because of ampere flow in the rotor bars. When peak magnetism in the rotor poles and stator poles occurs at nearly the same time, the result is very high starting torque (Fig. 5.11).

Ampere flow is restricted by the small size of the rotor bars. Power transformed into the rotor bars comes from the stator winding. (Ampere flow in the stator is normally high when starting.)

Large Rotor Bars

As with small rotor bars, the power transformed into large rotor bars comes from the stator pole's rotating magnetic field. (Large bars can carry higher current than small bars.) The current flow forms poles in the rotor iron. Large rotor bars (because of their shape) are surrounded by a lot of iron. Magnetizing the surrounding iron causes a delay of current flow in the bars (inductance). This delay in current flow (in the rotor bars) causes peak magnetism to occur later in the rotor poles than in the stator poles. The magnetic power of the stator poles will have peaked (and decayed to a lower value) when the magnetic field of the rotor reaches its peak power. The result is a poor phase angle.

The torque-producing magnetic interaction of the two windings is highest when both windings reach peak magnetism at nearly the same time. Low starting torque is the result of out-of-step peak magnetism, not the number of amperes in the squirrel cage winding.

Code Letter

The code letter (found on the motor's nameplate) determines a single-phase or three-phase motor's inrush amperes. Use the code letter and the following formulas to select the right overload protection:

Code letter value × hp × 1000 ÷ voltage for single-phase

Code letter value × hp × 577 ÷ voltage for three-phase (1000 × 1.73 = 577)

For 40 horsepower, 440 volts, three-phase, code F (5.3 is midrange),
5.3 x 40 × 577 ÷ 440 = 278 inrush amperes

Three-phase motor controllers are horsepower rated. The overload protection size is selected to fit each individual motor.

Torque and Speed

Torque is twisting force, measured in foot-pounds, inch-pounds, and inch-ounces. The formula combines torque and speed (RPM) to find horsepower:

torque × RPM ÷ 5250 = hp

Torque and RPM terms are explained in Fig. 5.12. Understanding these terms is essential when fitting a motor to a load.

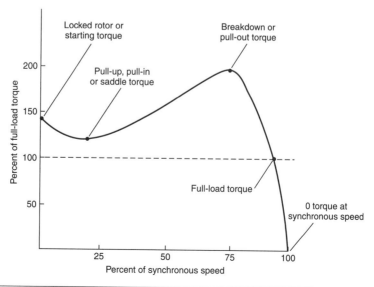

FIGURE 5.12   Torque terms from 0 RPM to synchronous speed.

Starting torque (locked rotor torque) is the amount of torque that is available to break the load away from a standstill position. (Breakaway torque is another name used for this torque value.)

Starting torque is a very important value when fitting a motor to a load. For example, an air compressor (starting against compression) requires very high starting torque.

Saddle torque is a dip in the motor's torque output that occurs between starting and pull-out (or breakdown) RPM. If the motor's (saddle) torque drops below the load torque demand, the motor will stop accelerating. The motor windings will overheat and fail (unless the overload protection trips).

Saddle torque is a problem found in two-pole motors that have a large-diameter bore. A special connection called the *interspersed connection* is used to minimize saddle torque. The problem occurs when a motor has been rewound and the connection is not used.

Breakdown (or pull-out) torque is the maximum possible torque that a motor can develop, and it can occur at any RPM between starting and full load.

If breakdown torque occurs close to nameplate RPM, just a few RPM less than nameplate RPM will cause the motor to overheat.

A small reduction in RPM, caused by an overload, results in a large increase in horsepower output (Fig. 5.13 design A). An increase in horsepower output is accompanied by an even larger increase in amperes (and heat).

Full-load torque is the maximum torque the motor can have continuously (without overheating). This torque occurs at the nameplate RPM.

The preceding torque values vary with rotor bar design. The motor's design letter—found on the motor's nameplate—identifies the torque characteristics of a motor.

Design Letter

A three-phase motor's design letter (found on the nameplate) is a very important factor in selecting a motor. Figure 5.13 shows the torque and speed characteristics of each letter.

Design A

The design A motor has large rotor bars, so it will demand high starting current. However, despite having high starting current, it has only moderate starting torque. Therefore, the motor's protection has to be adjusted accordingly.

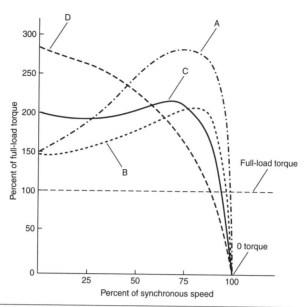

FIGURE 5.13   A design letter is given to a rotor. The shape, size, and location of the rotor bars affect the design letter.

This motor has very high torque near its rated RPM. Its large rotor bars keep the rotor's full-load speed close to the synchronous speed. (This can be compared to a paddle wheel in a stream of water; with large paddles the wheel slips very little and rotates at close to the water's speed.) The design A motor will keep its load at nearly constant speed.

The design A motor will overheat even if it's overloaded only a few RPM (below its nameplate rating). Higher torque (at lower speed) results in a large increase in horsepower output and ampere demand.

Some efficiency is lost with rotor slip. The torque curve of design A is similar to that of a high-efficiency motor.

Design B

Design B is a standard industrial motor. It has average starting torque and moderate starting current. This motor has a lower full-load RPM than that of the design A or the high-efficiency motor. Most design B rotors use low-cost cast aluminum.

Design C

The design C motor has higher starting torque and slightly greater slip than the design B motor. Its rotor bars are two sizes, linked together. A small bar

FIGURE 5.14 The rotor bar used in a design C rotor has two shapes that are linked.

near the rotor's surface is linked to a larger bar deeper in the iron (Fig. 5.14). The small part of the bar gives the rotor high starting torque. The deeper part will reduce the slip. Starting current is moderate, limited by the size of the small bar. The bars are usually made of copper or brass (making the motor more costly). They are more likely to crack because of their shape and the type of metal used.

Design D

Design D is for loads requiring extremely high starting torque but not high torque near rated speed. This design is suited for a heavy flywheel load or for a centrifuge load. This motor can power a heavy flywheel load up to the nameplate RPM without developing excessive heat. (Its small rotor bars limit the starting current.)

A metal shear is an example of a load that stores energy in the weight of its moving flywheel. The shear slows its motor's RPM below the nameplate value each time it cuts metal. Because of the small rotor bars, the torque and (stator winding) amperes aren't excessively high when the motor's speed is lowered.

The design D motor can handle wide RPM swings without developing excess heat in its stator windings.

Design Letter Summary

Industry presents a wide range of speed and torque demands for the three-phase induction motor. Four-pole motors are the most common, followed by two-pole, then six-pole motors. There are many different torque demands, from zero to full-load speed. Some loads require very high starting (breakaway) torque. Other loads require little starting torque, but demand higher torque as the speed increases. The design letter helps in selecting the right motor.

The design A and the high-efficiency motors develop high torque just a few RPM slower than their rated RPM. Because of this characteristic, they are the choice for loads that require a fairly constant speed.

Liquid-moving pumps and air-moving equipment have similar loading characteristics. If the RPM is increased a small amount over the rating, the horsepower demand increases to the cube of the RPM change. (The RPM of the replacement motor has to be the same as the original for this type of load.) Most three-phase induction motors operate very close to their nameplate RPM when rated load is applied. Less than rated load allows the shaft to turn at a slightly higher RPM than its nameplate value (slip can vary from 0.3 to 5 percent). Standard three-phase induction motors under 10 horsepower have greater slip than those over 10 horsepower. Slip for standard motors above 50 horsepower is 1 percent or less.

Introduction to Three-Phase Motor Types

The nine-lead three-phase induction motor is the most popular motor for industrial use. It ranges from fractional to hundreds of horsepower. The 12-lead induction motor is very similar to the nine-lead motor. Other induction motors covered are multispeed, multimode, and wound rotor motors. The synchronous motor and the brushless DC motors are also covered.

The Nine-Lead Induction Motor

The nine-lead three-phase induction motor can be used on two voltages. It will be connected as either wye or delta. Figure 5.15 is a schematic of a nine-lead wye connection. The delta connection in Fig. 5.16 is also designed for both voltages.

An ohmmeter can be used to determine whether the motor has a wye or delta connection (if there isn't a connection nameplate). As the schematics show, a nine-lead wye connection has three circuits with two leads and one circuit with three leads. The delta connection has three circuits with three leads each.

Each phase (in all nine-lead dual-voltage three-phase motors) is divided into two equal circuits. Regardless of the number of poles, one-half are in each circuit. The nine-lead motor has a 2-to-1 difference in its voltage rating.

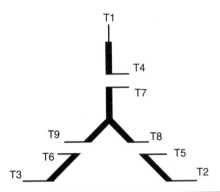

FIGURE 5.15 The most common connection found in small motors in the nine-lead wye connection.

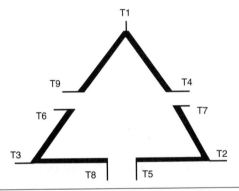

FIGURE 5.16 The nine-lead delta connection is usually found in motors of 10 horsepower and above.

Each pole has the same voltage across it, whether the motor is connected high or low voltage.

A telephone keypad (Fig. 5.17) simplifies three-phase connections. Exactly one-half of each phase is connected between two keys. For example the A phase, one half of its winding is connected between keys T1 and T4, and the other half between key T7 and the blank key below it.

The keypad can be used as a reference for wye, delta, 12-lead, 10-lead, and wye-delta connections. The keys have windings connected to them as shown in Fig. 5.17.

A wye-connected motor will have the bottom three keys tied together (internally) to form the wye connection.

For the 10-lead connection, lead number T10 is connected to the internal wye.

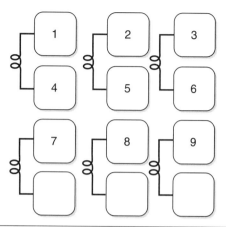

FIGURE 5.17   Internal winding-to-lead connections for three-phase motors using a telephone keypad.

If the motor has a delta connection, the bottom three keys (from left to right) are numbered 2, 3, and 1 and are connected, internally, to their respective lead numbers.

For the 12-lead connection, the bottom three keys, from left to right, are numbered T10, T11, T12, as shown in Fig. 5.28, later in the chapter.

The keys 1 through 6 only are used in the wye-delta connection. A complete phase is connected between each pair as follows: T1 and T4 contain the A phase, T2 and T5 the C phase, and T3 and T6 the B phase. The wye-delta connection sequence is shown in Fig. 5.54, later in the chapter.

Nine-Lead Wye Connection

A nine-lead wye-connected motor will have one or more wyes connected internally. The windings connected to leads T7, T8, and T9 end at the internal wye(s). Leads T4, T5, and T6 are connected together to form the external wye for the low-voltage connection (Fig. 5.18). Leads T1, T2, and T3 are always connected to lines.

These are two separate wye circuits with exactly the same internal data. There isn't an internal connection between the two circuits. One-half of each phase is in each circuit. The data are the same in circuits T1 and T4, T2 and T5, and T3 and T6.

For low voltage (two wye), connect T1 and T7 to line 1, T2 and T8 to line 2, and T3 and T9 to line 3 (Fig. 5.19). Leads T4, T5, and T6 are connected together and insulated.

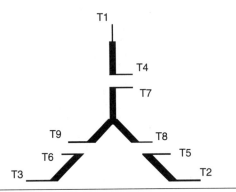

FIGURE 5.18 The nine-lead wye schematic.

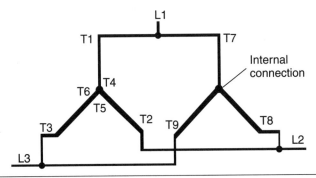

FIGURE 5.19 The low-voltage (two-wye) connection.

Both wye- and delta-connected motors combine the same lead numbers for high voltage. In both connections, each phase is connected in series for high voltage. In the wye connection (Fig. 5.20), a circuit from line 1 to line 2 goes through two phases.

Nine-Lead Multiple-Wye Connection with T10

The internal wye sometimes has lead T10 connected to it. This lead is connected to T4, T5, and T6 when the motor is connected low voltage (Fig. 5.21). When the motor is connected high voltage, T10 is insulated and not used.

The purpose of T10 is to balance the voltage (and the resulting current) between the wye connections. Ideally, wye connections would have no voltage or current difference between them. (In reality, there is always some unbalance.) The most common reason for the unbalance is a slight difference in air gap around the rotor. Connecting T10 as shown in Fig. 5.22

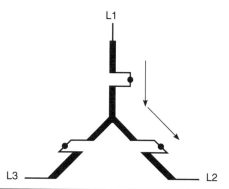

FIGURE 5.20   The high-voltage wye connection showing where the current flows from line 1 to line 2.

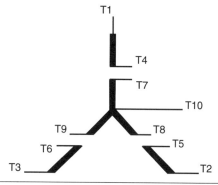

FIGURE 5.21   The 10-lead dual-voltage motor.

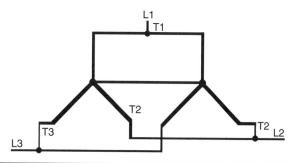

FIGURE 5.22   Multiple wyes are often connected together internally.

equalizes the voltage and current between the two connections. The result is a better magnetic balance between the stator poles.

For example, a 30-horsepower, two-pole, multiple-wye-connected motor that powered a metal polisher had to be rewound. When the motor was rewound and reinstalled, a pattern developed on the metal, instead of a mirror finish. The wye connections were joined, and the pattern disappeared.

Larger nine-lead motors have multiple wyes connected internally. The wyes are usually connected (internally) to a common wire, as shown in Fig. 5.22.

Thermal Protector for the Nine-Lead Wye-Connected Motor

A thermal protector for the dual-voltage (230/460-volt) wye-connected three-phase motor is shown in Fig. 5.23.

This protection is designed for small wye-connected motors. (The delta connection is seldom used in motors under 5 horsepower.)

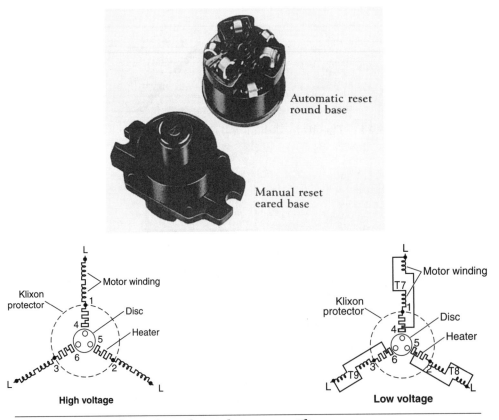

FIGURE 5.23 A three-phase thermal protector. *Klixon.*

In Fig. 5.23, there are three current-monitoring heat elements. High current through these heat elements will trip open the protector's contacts. They are located between 1 and 4, 2 and 5, and 3 and 6.

The motor's internal wye is connected permanently to terminals 1, 2, and 3 of the thermal protector. These leads are the ends of the phases. (In a 12-lead motor, they are labeled T10, T11, and T12.)

P4, P5, and P6 bypass the heat elements. When the motor is connected low voltage, they are connected to the motor's corresponding T leads (P4 to T4, P5 to T5, and P6 to T6). If connected high voltage, P4, P5, and P6 aren't used and are separately insulated.

If line 1, 2, or 3 becomes open, the motor operates on single phase. (The term *single-phased* describes this condition.) When a motor is single-phased, its power drops to approximately one-half of its horsepower rating. The amperes increase through two of the heat elements, causing the protector's contacts to open and disconnect all the motor's circuits. The thermal protector uses the amperes of T7, T8, and T9 to protect the motor, whether the connection is high or low voltage.

Nine-Lead Delta Connection

In the delta connection, T1, T2, and T3 are the points of the delta (Fig. 5.24). They contain the end half of one phase and the start half of the next phase. In the start half of phase A, lead T1 is connected to (the end) half of phase B. One-half of phase A starts at T1 and ends at T4. Lead T7 completes phase A,

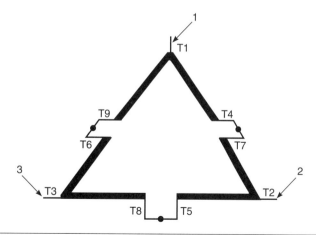

FIGURE 5.24   The points of the delta connection are numbered T1, T2, and T3 in the nine-lead schematic.

ending at T2. Phase C starts at T2 and ends at T5. Lead T8 completes phase C, ending at T3. Phase B starts at T3 and ends at T6. Lead T9 completes phase B, ending at T1.

When the delta is connected low voltage, each line is connected to three motor leads. Line 1 is connected to T1, T6, and T7; line 2 to T2, T4, and T8; and line 3 to T3, T5, and T9. Figure 5.25 shows how the circuits of each phase are connected in parallel.

Multiple-Delta Connection

The delta connection is found on motors above 5 horsepower. Larger nine-lead motors may have multiple deltas, connected internally. The external lead numbers are the same, and they are connected as described in Fig. 5.26.

Both delta- and wye-connected motors combine the same lead numbers for high voltage. Each phase is connected series in both connections. As shown in Fig. 5.27, each phase is connected across the line.

Twelve-Lead Connection

The 12-lead connection gives a motor greater versatility. A motor with this connection can be used on four different voltages.

In Fig. 5.28, the end of each phase is accessible by leads T10, T11, and T12. The rest of the leads (1 through 9) use the same lead number combinations as the nine-lead connection. Figures 5.29 through 5.32 show four different voltage connections that can be used on the 12-lead motor.

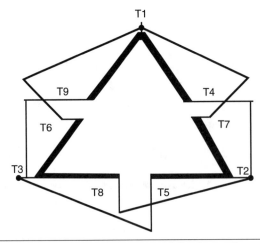

FIGURE 5.25 The delta connection connected for low voltage.

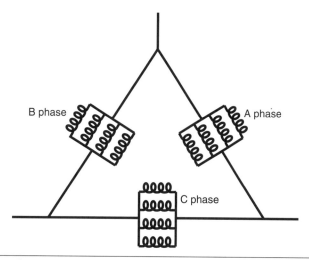

FIGURE 5.26   The multiple-delta connection is found in larger motors.

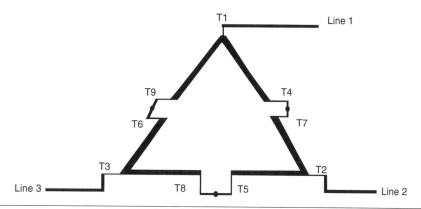

FIGURE 5.27   The delta connection connected for high voltage.

The lowest voltage connection is two delta (Fig. 5.29). The next-higher voltage connection is two wye (Fig. 5.30). (The two-delta voltage multiplied by the square root of 3 [1.73] equals the correct two-wye voltage.) The next-higher voltage connection is one delta (Fig. 5.31). This voltage is twice the two-delta voltage.

The highest voltage is one wye (Fig. 5.32) (twice the two-wye voltage). The voltage across each pole is the same on all four voltages.

The 12-lead connection can be connected part-winding start, using half-winding or two-thirds-winding start methods. The wye-start, delta-run, one-delta start, and two-wye run starting methods also work. (These start methods are discussed later under their separate headings.)

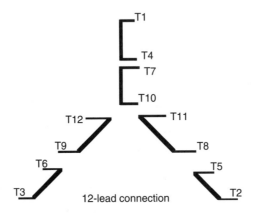

FIGURE 5.28   The 12-lead connection has the ends of each phase available for both wye and delta connections. The rest of the lead numbers are the same as those for a nine-lead wye or a delta connection.

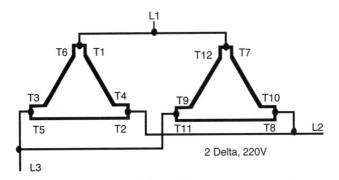

FIGURE 5.29   Two delta is the lowest voltage connection for the 12-lead motor.

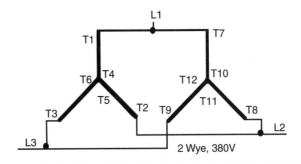

FIGURE 5.30   Voltage of the two-wye connection is 1.73 times that of the two-delta connection.

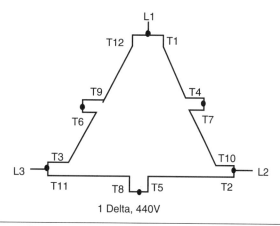

FIGURE 5.31   One-delta voltage is twice the value of two-delta voltage.

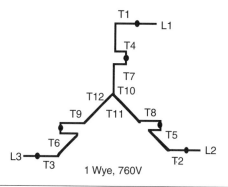

FIGURE 5.32   The highest voltage for the 12-lead connection is one wye—twice that of the two-wye connection.

Multispeed Motors

Multispeed motors covered in this section have one or more windings. The selected speed operates as a single-speed induction motor. (There is a small amount of slip, which is governed by the rotor design.) They are physically large for their rated horsepower when compared to single-speed induction motors.

Single-Winding Two-Speed Motors

Three types of single-winding two-speed motors are covered in this segment: constant horsepower, constant torque, and variable torque. All have a 2-to-1 speed ratio.

The two-speed windings have six leads that are switched and combined by a controller. The constant-horsepower and constant-torque motors use a one-delta and a two-wye combination. The variable-torque motor uses a one-wye and a two-wye combination.

The two-wye connection allows more current to flow than the one-delta or one-wye connection. This connection creates high torque (low speed) in the constant-horsepower motor and maintains the same torque (for both speeds) in the constant-torque motor.

The two-speed single-winding motor has a unique coil and pole arrangement. The pole width and spacing allow the motor to be switched from salient pole to consequent pole, and vice versa. Like a DC field pole, the salient pole is a wound pole. When the salient poles are connected so that the adjacent poles are of opposite polarity, the motor is in high speed.

Connecting all the poles of the same polarity creates the consequent-pole winding. Each pole's width is narrower than normal, leaving a space (iron) between the poles. This iron becomes a consequent pole. As a result, the number of poles is doubled. Doubling the poles cuts the speed in half.

The consequent-pole (low-speed) connection has a much lower power factor than the salient pole connection. It's common for the consequent-pole connection to draw full (or over) nameplate amperes with no load.

Constant-Horsepower Motor

The constant-horsepower motor has a single horsepower rating on its nameplate, with the same horsepower rating for both speeds. It's connected two wye for low speed and one delta for high speed (Fig. 5.33).

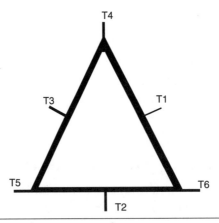

FIGURE 5.33   Schematic for a constant-horsepower motor.

If a constant-horsepower motor is connected low speed (two wye) and can lift 550 pounds 1 foot in 1 second, it will produce 1 horsepower. Connected high speed, the one-delta connection doesn't allow as much current to flow (as the two-wye connection). The motor now can lift only 275 pounds, but lifts it 2 feet in 1 second, maintaining 1 horsepower.

The constant-horsepower motor has high torque on low speed and less torque on high speed. It is ideal for a drill press application. A large drill bit needs high torque and should turn slowly. A small drill bit doesn't require much torque, and normally it is turned at high speed.

Constant-Torque Motor

The constant-torque motor has two horsepower ratings on its nameplate. The horsepower rating of its high speed is twice that of its low speed.

The connection is one delta for low speed and two wye for high speed. If a constant-torque motor is connected low speed (one delta) and can lift 550 pounds 1 foot in 1 second, it is producing 1 horsepower. When connected high speed, the two-wye connection allows more current to flow. More current gives the motor the ability to maintain the same torque, but at twice the RPM of low speed. At high speed, the motor can lift 550 pounds 2 feet in 1 second, producing 2 horsepower.

The constant-torque capability makes the constant-torque motor much more applicable than the constant-horsepower motor.

Variable-Torque Motor

The variable-torque motor has two horsepower ratings on its nameplate. The low-speed horsepower rating is one-fourth that of the high-speed horsepower rating. The variable-torque connection (Fig. 5.34) is one wye on low speed and two wye on high speed. The one-wye (low-speed) connection has one-fourth the horsepower that the two-wye (high-speed) connection has. The horsepower/speed combination is very compatible with the horsepower demands of a fan load.

A 50-horsepower, four- and eight-pole variable-torque motor is an example. Synchronous speeds are used instead of actual nameplate speeds (that involve slip). On high speed (1800 RPM), the motor develops 50 horsepower and is connected two wye. The low-speed (one-wye) connection will develop 12.5 horsepower (0.25 × 50 horsepower) at 900 RPM.

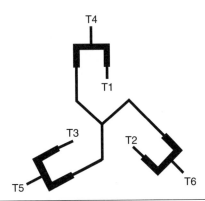

FIGURE 5.34 Schematic for a variable-torque motor. Heavy lines represent phase winding location relative to lead position. This schematic does not show true angle relationship.

When the speed of a fan is changed, the horsepower demand is not in proportion to the speed change. The horsepower must be changed to the cube of the ratio of the speed change. For example,

1800 ÷ 900 = 2 (ratio)   The ratio cubed (or 2 × 2 × 2) = 8

A fan requiring 50 horsepower at 1800 RPM would need only 6.25 horsepower at 900 RPM (50 ÷ 8 = 6.25 horsepower).

Any load requiring this type of horsepower/speed change ratio has a good potential for considerable savings in power cost. If the motor's speed can be lowered even a small amount and still meet the output demand, the power consumed will be reduced substantially. Air, liquid, conveyer belt, and auger loads all have similar speed/load power ratios, thus the potential for power cost savings.

Multispeed Multiwinding Motor

The multispeed multiwinding motor is usually constant torque or constant horsepower. The speed ratio for each winding is 2 to 1. Several of these windings are combined in some multispeed drill press motors.

If a multispeed motor has more than one winding, the higher-speed winding will have higher lead numbers. For example, the lowest speed will have numbers T1 through T7, the higher speed T11 through T17, and T21 through T27 on the next-higher speed winding.

Two-Speed Two-Winding Motor

A two-speed motor with a speed ratio that isn't 2 to 1—for example, 1200 RPM/1800 RPM—will usually have two single-speed windings. The lead numbers are T1, T2, and T3 for the low speed and T11, T12, and T13 for the high speed.

There's no internal connection between them, and they should be tested as two different motors.

If one or both windings are delta connected, the internal connection (described in Chapter 7, Fig. 7.4 under "Circulating Current in Multiwinding Multispeed Motors") must be used. A rewound motor draws excessively high amperes if this connection isn't used.

Multimode Nine-Lead Three-Phase Motors

Multimode motors are primarily used on oil well pumps. They must have high-slip rotors because of the uneven torque demand unique to oil well pumps. They also have special internal connections that are unlike those in any standard three-phase motor.

The external connections of the multimode motor are switched (by a control) to change the motor's torque without changing its speed. The control changes the motor's torque automatically when it senses a prolonged load change. The lower-torque mode reduces the amount of power consumed by the motor.

These motors are powered by either generators or long power lines, and they operate in remote locations. Both power sources benefit from having amperes as low as possible.

The multimode motor could be used on commercial air compressors. A commercial air compressor intermittently operates unloaded. During this time the standard motor consumes power and has a lower power factor. Replacing the standard motor with a multimode motor will reduce power cost while maintaining a good power factor during the unloaded part of the cycle.

Two types of multimode designs are covered in this section: the triple-mode design and the quadruple-mode design.

Comparing Multimode Motors to Standard Motors

Multimode motors have nine leads. Like the dual-voltage nine-lead delta connection, they have three sets of three-lead circuits. Unlike the delta connection, each three-lead circuit contains a complete phase (Fig. 5.35).

The 12-lead motor and the multimode motor both have leads that access both ends of each phase. Comparing the two in Fig. 5.36, the multimode motor's A phase has T7 located at the end of the phase instead of T10. The A phase of the multimode motor is an unbroken circuit with T4 located somewhere between T1 and T7 (depending on the number of torque modes).

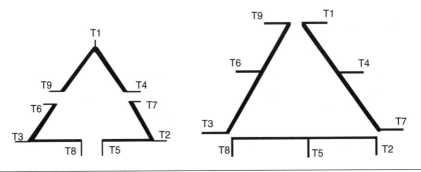

FIGURE 5.35 Comparison of the nine-lead delta schematic with a nine-lead multimode schematic.

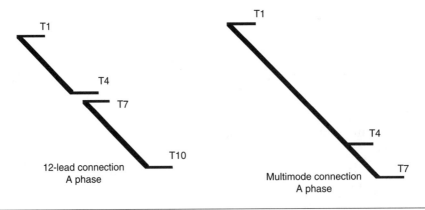

FIGURE 5.36 Comparing the A phase of a 12-lead motor to the A phase of a multimode motor.

Triple-Mode, Nine-Lead Three-Phase Motor

The triple-mode motor has a lead at the center of each phase. Its data are equal from the center lead to each end of the phase.

Figure 5.37 shows the connections for high, medium, and low torque. The delta connection (Fig. 5.37a) has the highest torque. Figure 5.37b is the wye-delta connection for medium torque, and Fig. 5.37c is the wye connection for low torque.

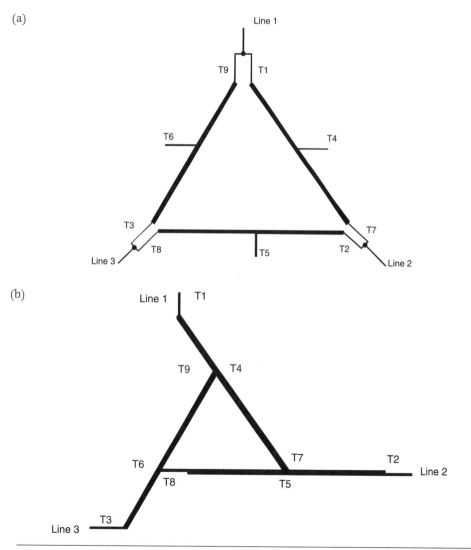

FIGURE 5.37   (a) The delta connection has the highest torque. (b) The wye-delta connection has medium torque.

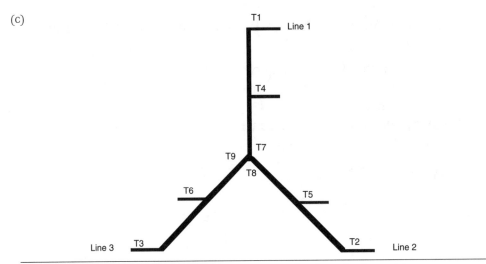

FIGURE 5.37 *(continued)* (c) The wye connection has the lowest torque.

Identifying the Triple-Mode, Nine-Lead Three-Phase Motor

The triple-mode motor has three separate circuits with three leads each (Fig. 5.37a). The center (of the A phase) lead is T4. Resistance will be the same from T4 to T1 as from T4 to T7. (The same is true of T5 and T6 and their respective leads, T8 and T9.)

Multimode motors have three numbers belonging to one phase (that is, T1, T4, and T7) in each three-lead circuit. The dual-voltage nine-lead delta connection has the numbers of two different phases in each three-lead circuit (that is, T1, T4, and T9).

If there are no lead numbers, find the center lead of one phase and connect the other two leads together. (See "Identifying Unmarked Leads in a Delta Connection" in Chapter 6.) Apply an intermittent DC voltage to this phase, and check the other two phases for voltage with a low-scale analog voltmeter. There should be little or no deflection. If it is a delta-connected nine-lead motor, the voltmeter will show a strong deflection between two of the three leads in the other circuits.

The multimode motor's nameplate will show three different horsepower ratings and three separate ampere values. There is one RPM value on the nameplate.

Quadruple-Mode, Nine-Lead Three-Phase Motor

The quadruple-mode nine-lead motor is the same as the triple-mode motor except for the location of the center lead. The center lead is located closer to the start of each phase; i.e., the A-phase lead T4 is closer to T1 than to T7. This gives the motor four torque modes (Fig. 5.38).

This motor has one RPM rating and four ampere ratings on its nameplate. If it's rewound, it is very important to replace the internal connections the same way they were in the original winding.

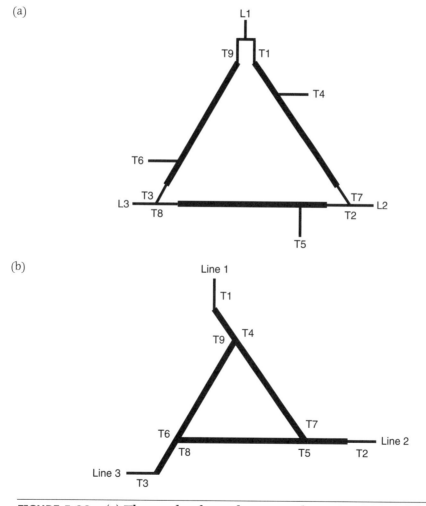

FIGURE 5.38   (a) The quadruple-mode motor schematic connected delta for the highest torque. (b) The quadruple-mode motor connected delta-wye will have medium torque.

(c)

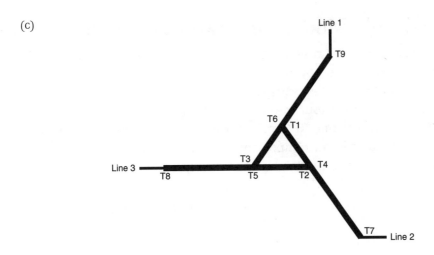

(d)

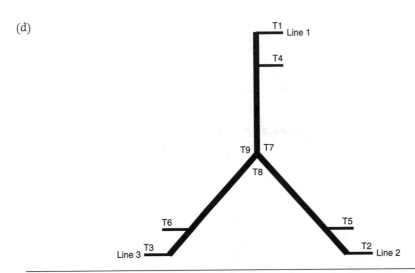

FIGURE 5.38 *(continued)* (c) The quadruple-mode motor connected wye-delta will have medium-low torque. (d) The quadruple-mode motor connected wye will have the lowest torque.

Figure 5.38a shows the delta connection, which has the highest torque. Figure 5.38b shows the delta-wye connection with medium torque. Figure 5.38c shows the wye-delta connection with medium-low torque. Figure 5.38d shows the wye connection, which has the lowest torque.

Identifying the Quadruple-Mode, Nine-Lead Three-Phase Motor

The same identification logic is used on this motor as described above for the triple-mode motor. The quadruple-mode motor's connection, however, is slightly different from that of the triple-mode motor.

This connection can be identified with an ohmmeter of the right size. There are three sets of three leads that light to each other. A-phase leads T1 to T4 have much less resistance than T4 to T7. The same is true of T2 to T5, T5 to T8, T3 to T6, and T6 to T9.

Synchronous Motor

Synchronous motors are used on loads that need constant speed. These motors range in size from small to many thousands of horsepower. Figures 5.39 to 5.42 are various pictures of the synchronous motor and its components.

FIGURE 5.39   The synchronous motor. *Smith Services.*

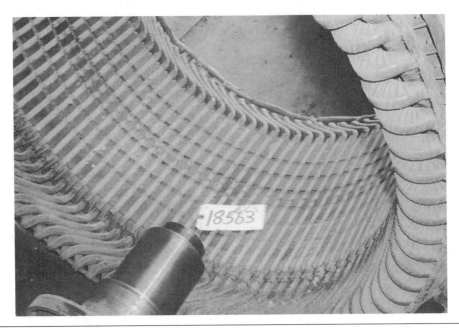

FIGURE 5.40   Stator of a synchronous motor. *Smith Services.*

DC field coil   Pole iron

FIGURE 5.41   Rotor of a synchronous motor. *Smith Services.*

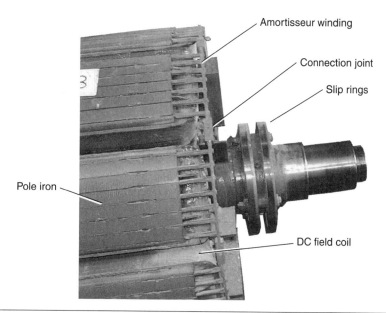

Amortisseur winding

Connection joint

Slip rings

Pole iron

DC field coil

FIGURE 5.42   Amortisseur winding (squirrel cage) embedded in the iron of the DC poles. *Smith Services.*

The synchronous motor doesn't have good starting torque and must start with very little load. Common applications are refrigeration compressors and air compressors, because these units can be started with no load and then activated after the motor is up to speed.

The synchronous motor is often used to correct the power factor in industry.

Components of the Synchronous Motor

The synchronous motor has a three-phase stator and frame, as an induction motor or the wound rotor motor (described later) has. The stator winding is either wye or delta connected (or with multiple wyes or deltas) internally. They are usually connected one voltage. Most operate on high voltage (2400 volts and higher) and have form-wound coils (Fig. 5.40).

The rotor of the synchronous motor has DC field poles mounted on it (Fig. 5.41). There are as many poles on the rotor as there are in one phase of the stator. The DC coils get their power from two slip rings mounted on the shaft.

The DC field of the synchronous motor is similar to the shunt field of a DC motor. (The resistance of the coils limits the amperes.) The coils (mounted on the rotor) are well cooled.

The pole iron for the DC poles is laminated, like the stator poles. It's fastened securely to a frame called a spider. The spider is mounted on the motor's shaft.

Each pole has a squirrel cage winding embedded in it above the DC coils (Fig. 5.42). This is called the amortisseur winding. The bars in the pole iron have both ends connected to sections of an end ring. The end ring sections and bars stay with each pole iron piece when the rotor is disassembled. Additional end ring sections join the pole end ring sections to form a complete circuit. The ring sections are lapped and bolted together securely, both mechanically and electrically. Like any squirrel cage winding, the circuit has low voltage, and carries very high amperes when starting. When the DC field is energized and the rotor is at synchronous speed, there is no voltage or current in the amortisseur winding.

Operation of the Synchronous Motor

When the stator is energized, the synchronous motor starts like an induction motor. The amortisseur winding (like a squirrel cage winding) forms poles that react to the rotating magnetic field of the stator.

The amortisseur winding isn't designed for long-term use (as is the case with the induction motor's squirrel cage winding). It gives the motor adequate torque to start a minimal load. A control energizes the rotor's DC field when the rotor reaches 95 to 98 percent of synchronous speed. The energized DC field quickly pulls the rotor up to synchronous speed. The motor is now ready for loading.

During startup, the lines of force from the stator's rotating magnetic field cut the DC field coils' conductors. At this time, the stator becomes the primary and the DC field becomes the secondary of an extremely high-voltage transformer. The stator coils have very high volts per turn, and very few turns. The DC field coils contain a large number of turns. When transformer theory is applied to this combination, the result is extremely high voltage in the DC field. The DC field coils are connected in series, so the voltage value of each coil adds to the next, resulting in very destructive high voltage at the slip rings.

A discharge resistor is connected across the DC field during startup (to lower the transformed high voltage). The discharge resistor affects the DC field in the same way that an overload affects a transformer. (When a transformer is overloaded, the secondary voltage drops.) The discharge resistor overloads the DC field's transformed voltage, keeping it at a safe value.

For an instant (when the control applies DC voltage to the DC field) DC flows through both the discharge resistor and the DC field. The discharge resistor is then disconnected.

It's very important that the DC field always have either DC power applied to it or the discharge resistor connected across it. A destructively high voltage will develop instantly if the DC field is open. This is one of the main causes of premature insulation failure in DC field coils.

The Synchronous Motor Control

The AC side of the synchronous motor control is similar to most motor controls and doesn't need explaining.

The DC side of the control is shown in Fig. 5.43. The amortisseur winding powers the rotor to 95 to 98 percent of synchronous speed (when AC is applied to the stator of the synchronous motor). As the rotor accelerates, a discharge resistor controls the voltage transformed to the DC field. An AC relay coil (no. 1) is connected across part of the discharge resistor and operates on the voltage drop across this part of the resistor.

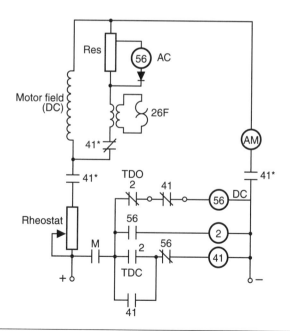

FIGURE 5.43   DC side of a synchronous motor control.

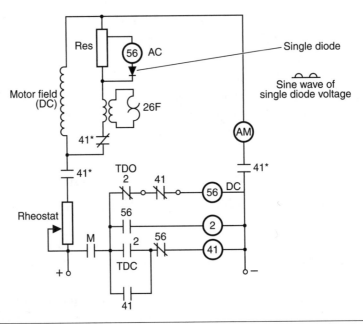

FIGURE 5.44 The single diode allows only half of the sine wave voltage to energize the control coil. The DC field and resistor are in a closed circuit during startup.

The voltage transformed to the rotor's DC field during startup is AC. The hertz (cycles per second) of the transformed voltage varies inversely with the rotor's RPM. (See "Amperes and Rotor Bar Design," earlier in the chapter.) A single diode is in series with coil 56. It will conduct only one-half of the sine wave. The half-wave pulse powers coil 56 (Fig. 5.44). As the rotor approaches 95 to 98 percent of synchronous speed (when DC should be applied), the half-wave pulses are down to about one pulse per second. The time delay between pulses is now so long that the coil can't hold the contacts. But the voltage is always high enough to keep the coil operating. (The time lag between pulses makes the coil release the contacts.)

Three sets of contacts are activated when coil 56 drops its load (Fig. 5.45). Two sets (normally open) connect DC power to the field. The third set (normally closed) functions after the other two sets close. For an instant, the resistor and the DC field have DC power applied to them. Then a normally closed contact opens and disconnects the resistor from the circuit.

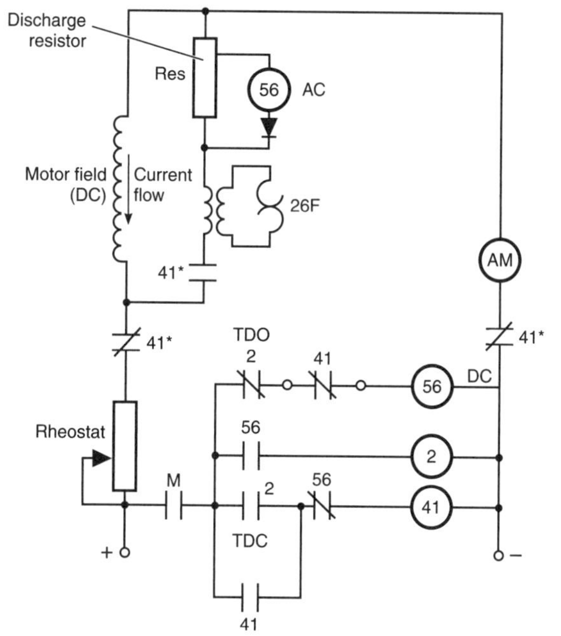

FIGURE 5.45   Current flow when motor is at full speed.

Power Factor Correction with a Synchronous Motor

The synchronous motor can be used to control the power factor of an industry's power supply by increasing or decreasing DC field amperes.

Unlike power factor–correcting capacitors (that correct only power factor), the synchronous motor corrects the power factor and pulls a load. It operates at constant speed. A load can vary from no load to full load (while not affecting the motor's RPM).

When no more power is applied to the DC field than it takes to pull the load, the AC power line magnetizes the iron of the AC stator. Magnetizing amperes (furnished by the power line) lower the line's power factor. (By increasing the amperes in the DC field, the DC field provides the magnetizing amperes.) Increasing DC amperes even more creates the same effect as with a capacitor (the current leads the voltage). Magnetizing amperes (furnished by the synchronous motor's leading current) raise the power factor of the AC supply circuit.

There are synchronous motors (with no shaft) that are used only for power factor correction. They are usually located where power factor correction is most beneficial to the building's circuitry.

The Brushless Synchronous Motor

The brushless synchronous motor operates the same way as the brush type. The main difference is the way the DC rotor is powered. Eliminating the brushes decreases maintenance of the machine.

The brushless synchronous motor has an AC alternator (or exciter) built into its frame. Instead of brushes carrying power to the DC rotor field, an AC alternator (exciter) furnishes the DC power. The exciter's stator field is DC and is controlled electronically by an external controller (sometimes called the black box). The black box controls the exciter and the output of the rotor components.

The receiver coils on the rotor cut the magnetic lines of force of the exciter's stator field. The power produced by the receiver coil is AC. A rectifier is mounted on the rotor shaft with the receiver coil. Control of the discharge resistor (also mounted on the motor's shaft) and the DC field is achieved with electronic components mounted on the rotor.

Three-Phase Alternator

The synchronous three-phase motor can be used as a three-phase alternator. The motor is rotated by a power source (diesel, turbine, or an electric motor), and DC is applied to the rotor's DC field. The DC field's flux cuts the stator windings and produces three-phase power.

The voltage (taken from the stator leads) can be varied by adjusting the amperes in the DC field.

Hertz is controlled by the speed of the rotor. If the alternator's power is added to an existing power source, the voltage polarity value and hertz must match those of the existing power source.

The amortisseur (squirrel cage) winding in the three-phase alternator has a dampening effect on speed change (hunting) in the machine when it is lightly loaded.

Wound Rotor Motor

The wound rotor motor is a variable-speed induction motor (Fig. 5.46). It was one of the first variable-speed three-phase motors to be developed. Sizes range from fractional to thousands of horsepower.

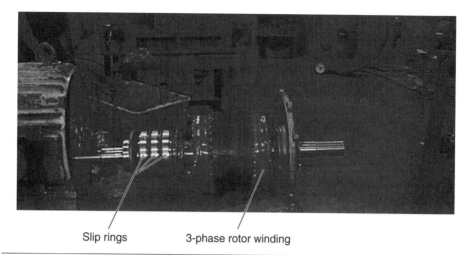

Slip rings   3-phase rotor winding

FIGURE 5.46   The wound rotor three-phase motor (disassembled). *Smith Services.*

Components of the Wound Rotor Motor

The wound rotor motor (Fig. 5.47) has a three-phase stator and frame similar to the induction motor described earlier. The stator winding is wye or delta connected (or with multiple wyes or deltas) (internally) and is usually connected one voltage.

The rotor has a three-phase winding similar to the stator winding (Fig. 5.48). (The pole spacing is identical to the stator pole spacing.) The rotor winding is also wye or delta connected (or with multiple wyes or deltas). Each phase is connected to a slip ring. The rings have brushes riding on them that are connected to leads M1, M2, and M3. These leads connect to a controller that controls the ampere flow in the rotor windings.

Operation of the Wound Rotor Motor

Three-phase voltage is transformed into the rotor winding when the stator winding is energized. This creates a three-phase voltage potential across the three slip rings. The voltage value depends on the turn ratio between the rotor winding poles and the stator poles. (Turn ratio is explained under "Transformer Function" in Chapter 3 and illustrated in Fig. 3.42.)

When the motor starts, the rotor control is set at its highest resistance value. The simplest explanation of the control is given in Fig. 5.49. This is a wye-connected set of resistors. (It's very important that resistance be absolutely equal between all three phases.) Figure 5.50 shows a more

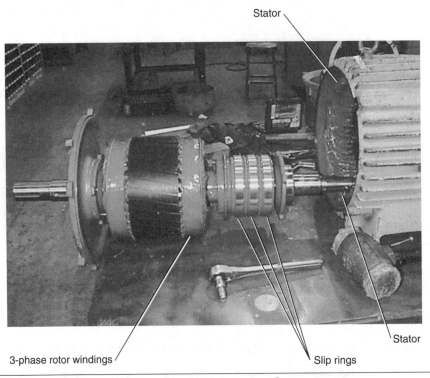

Stator

Stator

3-phase rotor windings

Slip rings

FIGURE 5.47 Stator of a wound rotor motor. *Smith Services.*

3-phase rotor winding
connected to slip rings

FIGURE 5.48 Rotor of a wound rotor motor. *Smith Services.*

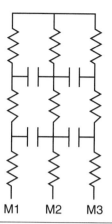

FIGURE 5.49   Control resistors for the wound rotor motor.

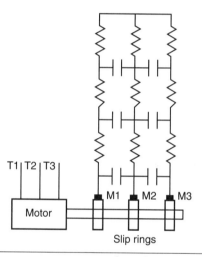

FIGURE 5.50   Wound rotor motor's control, with contacts for bypassing
each resistor.

detailed control with resistors (and contacts that bypass them as more power
is needed).

The control allows a low amount of current to flow in the rotor winding
poles when the motor starts. The stator and rotor windings have a good
phase angle (see "Phase Angle" earlier in the chapter) and good starting
torque with low current. As the rotor accelerates, the controller decreases its
resistance until the rotor reaches full speed. At full speed, the slip rings are
shorted together, and the rotor operates as a squirrel cage rotor.

If resistance is cut out too fast, the rotor windings develop high current
and high inductance. (Inductance delays the time that current flows.) When

peak magnetic power is reached in the rotor poles, the stator pole's magnetic power has already peaked and is dropping to a lower value. The rotor pole's peak power is now out of step with the stator pole's peak power—a bad phase angle. The result is torque loss. The motor will still bring the load up to speed, but the result is excessive heating and needless power loss.

Even on the lowest speed setting, a wound rotor motor will accelerate to near synchronous speed if there is no load. From zero to full RPM, the control and the load regulate a rotor's speed.

Until the motor reaches full speed (when the slip rings are shorted), the motor has low efficiency. To increase efficiency, the power (that is normally wasted as heat with resistors) can be converted to three-phase power, with a regenerative electronic control. This power is returned to the line, lowering the motor's operating cost.

Liquid Used for Speed Control

Another type of rotor current control uses liquid as its resistance. Three plates (connected to the slip ring leads) are lowered into a liquid. Current flow increases as the plates are lowered. The current control is much smoother than one with resistors and contacts.

Reversing the Wound Rotor Motor

Interchanging any two stator leads reverses the wound rotor motor. Interchanging the rotor leads won't reverse this motor, because the stator poles determine the polarity of the rotor poles.

The wound rotor motor should not be plugged (reversed) while running. Plugging transforms a very high voltage into the rotor winding. Voltage produced (by plugging) in the wound rotor winding is too high for the circuit's insulation.

The highest voltage will be at the end of the phases (leads). (The rotor circuit's weakest insulation is between the exposed metal of the slip rings and the shaft.) If one slip ring is grounded to the shaft, the insulation thickness between the rings is cut in half. Arcing will occur between one or both of the other two slip rings and the shaft. Brush dust contamination (in the slip ring area) increases the chances of an arc forming.

Plugging has resulted in rotor winding leads being blown off the slip rings. Voltage dramatically increases in the rotor winding, and has caused a flashover between the slip rings.

Wound Rotor Motor Used as a Variable-Voltage Transformer

A wound rotor can be used as a three-phase variable-voltage source. A geared device controls the movement of the motor's shaft. The stator winding acts as the primary of the transformer, and the rotor winding is the secondary. The slip rings aren't used.

When the rotor is moved, the voltage transformed into the rotor poles varies as their alignment with the stator poles changes. Total pole-to-pole alignment produces full voltage output. When the rotor poles are moved 90 electrical degrees out of alignment (one-half pole), the voltage output is at its lowest value.

The device that controls the rotor's movement has to be strong enough to withstand the motor's full torque. Torque develops when amperes flow in the rotor windings. The rotor windings are usually redesigned and rewound to produce the desired voltage.

Wound Rotor Motor Used as a Hertz Changer

Before the development of electronic hertz controls, a specially modified wound rotor motor was used to produce 120-Hz power. The power was achieved by energizing the wound rotor motor's stator and driving its shaft against its rotation with another motor. The power was tapped off the slip rings of the wound rotor motor. The quality of power produced by this method is better than today's electronically produced 120-Hz power.

The Brushless DC Motor

The brushless DC motor is a type of three-phase synchronous motor (Fig. 5.51) with permanent magnets on its rotor. (Most synchronous motors have coils on their rotors, powered by DC.)

Like the synchronous motor, the brushless DC motor has absolute speed regulation. The speed control is exact (± 0). Unlike the synchronous motor, this motor doesn't shut down if it's overloaded. When there is an overload, the controller will automatically adjust the hertz to a different RPM.

The power range of brushless DC motors is from $1/_4$ to 400 horsepower. They have NEMA standard frame sizes and enclosures, and the controllers operate on either 50- or 60-Hz power sources.

The brushless DC motor control can include dynamic (regenerative) braking to slow a momentum load. It has full-load torque at zero speed,

FIGURE 5.51 Cut-away of brushless DC motor. *Powertec.*

for use as a brake or as a servomotor. Its acceleration and deceleration capabilities are typically 0.05 to 90 seconds, and they are separately adjustable in the controller. It has electronically controlled plugging capability (reversing) at any speed. The motor's current and voltage are electronically limited when the motor is plugged. This decreases high-ampere stress that plugging typically causes in other types of motors.

The brushless DC motor can withstand a 150 percent overload for 1 minute. It's protected with instantaneous trip action when the current exceeds 250 percent.

It has nearly unity power factor at all speeds and loads. Magnetizing amperes are limited to the stator. The rotor doesn't require any magnetizing current since the rotor's magnetic field is supplied by a permanent magnet. (The squirrel cage rotor requires continuous magnetizing.)

Components of Brushless DC Motors

The brushless DC motor has a three-phase stator winding. The winding is single-voltage and is operated by the motor's controller. It can't operate without its controller.

The rotor has permanent magnets epoxied to its shaft. The magnets are reinforced with several wraps of high-strength fiberglass and epoxy tape.

Hall effect sensors are connected to the motor's end bracket. A multipole magnet wheel is attached to the rotor shaft. The Hall sensor assembly has five to seven wires, which go to the controller in a separate conduit from the stator winding circuit. (The data carried by the control wires would be corrupted if they were in the same conduit as the stator leads.)

All brushless DC motors have ball or roller bearings. (It's very important to monitor the condition of the bearings closely.) If a bearing breaks down, the rotor will be destroyed. Bearings in a brushless DC motor last longer than those in most other types of motors (because of the low shaft temperature). Permanent magnets eliminate the heat that results from a squirrel cage's (magnetizing) winding current.

Operation of the Brushless DC Motor

The brushless DC motor operates from a variable-voltage variable-hertz speed controller. A Hall effect sensor (on the motor shaft) tells the controller the exact position of the magnets (relative to the stator poles). The rotating magnetic field is established when the controller alternately powers the stator's three phases. Two phases on and one phase off create a rotating magnetic field.

Three-Phase Servo Motors

Some servo motors operate much the same as the brushless DC motor. They have feedback devices called resolvers, encoders, tachometer generators, and Hall effect sensors, in combinations of two or more.

The servo motor's controller tracks the rotor's permanent magnets. It energizes the proper stator windings to position the shaft exactly as needed. It can also control the exact speed of the motor.

To determine if a motor has permanent magnets, short two of its leads together and turn the shaft. Permanent magnets will cause resistance to rotation.

If the motor is disassembled, it is very important to put reference marks on the shaft and end bell. Marking the location of components simplifies reassembling.

If the alignment is not recorded, energize two winding leads with low-voltage DC and controlled current (5 amperes or less). This will lock the rotor and provide a reference point for positioning the components. Most servo controls compensate for some misalignment. The number of poles can also be determined this way.

These tiny motors and their controls will probably become throw-aways like other small motors.

Identifying the Three-Phase Induction Motor

Three-phase induction motors have many different connections. Identification can be difficult, especially if there is no nameplate. Most of the connections and their function and application will be identified in this section.

Troubleshooting procedures found in Chapter 6 can be used on all the motors described in this section.

The number of leads a motor has can help identify its connection. The number can be from 3 to 18 or more. Most are divisible by 3, with the most common being the nine-lead dual-voltage motor.

Six-lead motors have many different variations and can be hard to identify. Wye-delta, constant-torque, constant-horsepower, and variable-torque motors are all six-lead motors (that use the numbers 1 through 6).

Connections encountered less frequently are two-speed one-winding, two-speed two-winding, part-winding start, wye-delta, 12-lead motors, dual-voltage multispeed, multimode, and European connections.

A motor with no nameplate presents a common identification problem. Sometimes a motor has been redesigned to a different horsepower, speed, or voltage (without the change being noted on the nameplate).

All motors should be tested using a current-limiting control, such as resistors. An unidentified motor should never have full power applied to it.

Nine-Lead Dual-Voltage Motors

A nine-lead dual-voltage motor will always be wye or delta connected.

The wye-connected nine-lead motor has three sets of two leads that light to each other and one set of three leads that light to each other.

The delta-connected nine-lead motor has three sets of three leads that light to each other. The multimode motor also has three sets of three leads that light to each other. There is, however, a big difference in the motor's internal connections. This is explained under "Multimode Nine-Lead Three-Phase Motors," earlier in the chapter. (This connection is found mainly in the oil field industry.)

Nine leads make it possible to operate a three-phase motor on either of two voltages. There is a 2-to-1 difference between the high- and low-voltage connections.

Nine-lead motors (both wye and delta) can also be connected part-winding start if they have the right style of winding. This is explained later in this section under "Concentric-Style Winding and Part-Winding Start Connection."

A variation of the nine-lead wye-connected motor involves a tenth lead. The tenth lead is used for joining the internal wye(s) when the wye-connected motor is connected low voltage.

Troubleshooting these motors is covered in Chapter 6.

Three-Lead Motors

Three-lead three-phase induction motors are designed for one voltage and one speed. (Larger motors usually have three leads, because they are seldom designed for dual voltage.) Nine-lead motors (when rewound) can be changed to three leads if they operate on only one voltage; thus there are fewer connection problems than with nine leads. (The motor's voltage rating should be stamped on the nameplate.)

Run the motor on low voltage (240 volts) with a current-limiting control if the voltage isn't known. If it starts with much less inrush current than expected, it's connected or designed for a higher voltage. The motor will have about one-fourth of its normal power and will start more slowly than normal.

It's hard to determine the right voltage for small motors. They start quickly on low voltage—although connected for high voltage—because of their lightweight rotors. No-load amperes will be extremely low when run on low voltage.

The three-lead motor's internal connection will be wye or delta. Both are tested with the same testing procedures found in Chapter 6.

Six-Lead Motors

When a motor has six leads, it can have many different connections. (Some examples are constant-torque, constant-horsepower, variable-torque, two-speed two-winding, part-winding start, and wye-delta connections.)

Six leads are sometimes used on single-speed induction motors, because there isn't room between the motor's core and shell for three larger leads. In this case, the leads of each phase should be secured in one lug so they stay together.

The data on the nameplate help to identify the six-lead motor's connection. The number of circuits between leads is also helpful. The motor's nameplate should have the horsepower, speed, and voltage. This information may be all that is needed to identify the motor's connection. If the nameplate doesn't give enough information, check the number of circuits between the leads. Circuit descriptions (in the following text) can then be used for identification.

A *two-speed one-winding* motor will have six leads that all show continuity to each other. They are numbered 1 through 6. Their nameplate will show different horsepower ratings (related to its design) but seldom will identify the motor's type. They all have a 2-to-1 speed rating and are oversized (for their horsepower).

Constant-torque motors have two horsepower ratings. (The high speed has twice as many horsepower as the low speed.) A *constant-horsepower* motor has the same horsepower rating for both speeds. *Variable-torque* motors have two horsepower ratings. (The low-speed horsepower is one-fourth that of the high-speed horsepower.)

The three connection schematics shown in Fig. 5.52 identify the types of connection used in these motors. (The constant-torque motor is the most common.)

Constant-Torque Connection

Connect the lead numbers (as shown on the constant-torque schematic for low speed) with T4, T5, and T6 open, and apply limited-current three-phase power. If the motor runs well, connect it high speed. If it operates satisfactorily, run the motor with full rated voltage on both speeds. Connected low speed, the motor may draw full or slightly above nameplate amperes. Connected high speed, it should draw approximately 50 percent less than the nameplate amperes, which would indicate that the motor is connected constant torque. This (constant-torque) motor has twice the horsepower on high speed that it has on low speed.

Variable-Torque Connection

Variable-torque and constant-torque motors use the same lead number combinations. If the motor's power is especially low (on low speed),

(a)

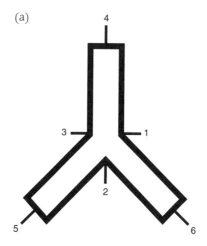

Low speed
L1–T1, L2–T2, L3–T3
T4, T5, T6 Open and Insulated

High speed
L1–T6, L2–T4, L3–T5
T1, T2, T3 Together

(b)

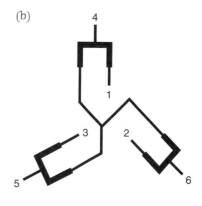

Low speed
L1–T1, L2–T2, L3–T3
T4, T5, T6 Open and Insulated

High speed
L1–T6, L2–T4, L3–T5
T1, T2, T3 Together

(c)

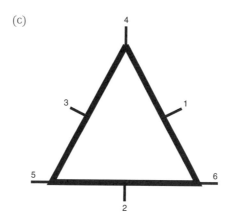

Low speed
L1–T1, L2–T2, L3–T3
T6, T4, T5 Together

High speed
L1–T6, L2–T4, L3–T5
T1, T2, T3 Open and Insulated

FIGURE 5.52   (a) The constant-torque connection. (b) The variable-torque connection. (c) The constant-horsepower connection.

the motor's connection is variable torque (Fig. 5.52b). The variable-torque motor's low-speed horsepower is one-fourth that of its high-speed horsepower. Load testing the motor is an easy way to tell the difference between the two connections.

Constant-Horsepower Connection

If the motor is designed to be constant horsepower (Fig. 5.52c), the low-speed constant-torque connection will cause it to draw very high amperes. If amperes are high, test the motor (with limited power) using the constant-horsepower connections.

If the motor connection has been positively identified and has very little power, it is connected (or redesigned) for higher voltage than is being applied. Test the motor (first, with limited power) using a higher voltage.

Two-Speed Two-Winding Connection

The two-speed two-winding motor has two sets of three leads that light to each other. The lead numbers are T1, T2, and T3 for low speed; T11, T12, and T13 for high speed. There may or may not be two horsepower ratings on its nameplate. If so, the low-speed horsepower will be less than that of the high speed.

The two-speed two-winding motor has the same number of circuits (two sets of three) as the part-winding start motor (to be covered next).

Part-Winding Start Connection

The part-winding start motor will have two circuits of three leads. One circuit is numbered T1, T2, and T3; the other, T7, T8, and T9. Each circuit contains a portion of a single-speed winding.

The most common connection is the half-winding start, using the two-wye connection. (One-half of the winding is in each circuit.)

Applying limited three-phase current to one circuit will identify this connection. If the motor starts and runs, do the same test with the other circuit. Both circuits should run at the same speed and amperes. The motor's power is roughly one-half with one circuit.

The motor may or may not be noisy (depending on the internal pole-to-pole connection), and the amperes will be low. Connect line 1 to T1 and T7, line 2 to T2 and T8, and line 3 to T3 and T9. The motor should start and run with full power. (The motor starts the load on one-half of its winding. After 2 or 3 seconds, the other half is energized.)

The inrush amperes are one-half the amps of across-the-line starting (using the complete winding). As the speed increases, the amperes in the first-stage winding drop. When the second-stage winding is energized, its ampere demand will be about the same as that of the first stage. With both stages energized, the total ampere demand is less than that of across-the-line start.

Part-winding starting is not recommended for a load that demands frequent or heavy prolonged starts. (Excessive heat will develop in the first-stage winding because it has full voltage applied to it.) If heat doesn't have time to dissipate (from frequent starts), the winding's insulation life is reduced. The wye-delta start method (discussed later in this section) is better for heavy prolonged starting.

Part-Winding Start of a Two-Pole Motor

Two pole motors can be part-winding started if they have the right internal connection. If T1, T2, and T3 are connected to the first pole of each phase, as illustrated in Fig. 5.53, the motor will start and run normally (when T7, T8, and T9 are energized). If T3 is connected to a pole in the other half of the winding, the motor may reverse itself when the second half is energized.

Concentric-Style Winding and Part-Winding Start Connection

The nine-lead motor (with a concentric-style winding) won't start on part of its winding. This problem occurs often when a lap wound motor, or a specially connected part-winding start motor, is replaced with a standard (concentric-wound) motor. A special connection (not used on standard nine-lead dual-voltage motors) has to be used.

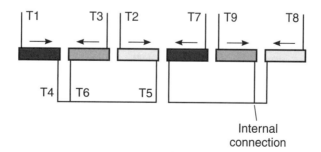

FIGURE 5.53   The internal connections for a two-pole motor that will work with a part-winding start control. The motor won't reverse when the second half is energized.

To keep up with competition and high-volume demand, most motor manufacturers use the concentric winding method. Concentric-shaped coils can be inserted into a motor's slots with a machine. (The lap winding has to be wound by hand.) Electric motor repair centers often redesign the concentric winding to lap winding.

Wye-Delta Connection

The wye-delta connection (Fig. 5.54) is used in Europe as a dual-voltage connection. In the United States, the connection is used primarily as a starting method.

A motor connected wye-delta will have three sets of two leads that light to each other (T1 to T4, T2 to T5, and T3 to T6). There will be two voltage ratings on the nameplate, with the higher voltage (wye) being 1.73 (the square root of 3) times the lower voltage (delta).

The operating voltage (for wye-delta starting) is that of the delta connection. If the delta connection is rated 220 volts, the wye connection voltage will be 380 volts (a difference of 160 volts). To produce full power, the volts per turn must be the same for both connections.

Verify the connection by connecting T4, T5, and T6 together; then apply limited three-phase current to T1, T2, and T3. (Use the voltage rated for the delta connection.) The motor should start slowly and have reduced power. Reconnect the motor (using the delta connection: line 1 connected to T1 and T6, line 2 to T2 and T4, line 3 to T3 and T5). If the voltage and connection are correct, the motor should run normally with full load.

Wye-Start, Delta-Run Starting Method

The wye-start, delta-run method of starting is similar to reduced-voltage starting. The wye connection is energized first with 220 volts (the delta

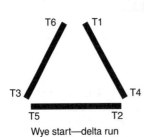
Wye start—delta run

Wye—delta dual voltage	L1	L2	L3	Together
Wye—high voltage (360v)	T1	T2	T3	T4, T5, T6
Delta—low voltage (220v)	T1, T6	T2, T4	T3, T5	

Wye start—delta run	L1	L2	L3	Together
Wye start	T1	T2	T3	T4, T5, T6
Delta run	T1, T6	T2, T4	T3, T5	

FIGURE 5.54   The wye-delta connection.

operating voltage). A few seconds later the control switches to the delta connection, and the motor runs at full power.

Using this method, the inrush amperes are much less than with across-the-line starting. This allows the motor to take more time to start a heavy load, without excessively overheating its winding.

The motor will have approximately one-third of its normal torque on the wye connection. The load must not require a breakaway torque higher than the wye connection can produce.

If a wye-delta-connected motor doesn't have enough breakaway torque, the one-delta start, two-wye run connection can be used. (The operating voltage must be for two wye.) The one-delta start, two-wye run connection (explained a bit later under "Twelve-Lead Single-Speed Motors") produces 40 percent more starting torque than the wye-delta method. Inrush amperes are higher than those of the wye-start, delta-run method.

Wye-Delta Induction Motor with Unknown Voltage

Sometimes a motor's nameplate is missing or doesn't show a voltage rating. If the coils are mush-wound lap or concentric, the motor's voltage may be rated anywhere from 208 to 760 volts (Fig. 5.55). (The mush winding is rarely used on voltage above 760 volts.) The coil wire is usually visible with the mush winding.

Form-wound coils are used in motors above 2400 volts. This type of winding is rarely found in motors rated 760 volts and less. The coils are each wrapped with insulation. A motor repair center or the motor's manufacturer should be consulted to determine the correct voltage.

Applying a voltage that is too high (for example, 460 volts to a 230-volt motor) will destroy a motor in a few seconds. Always use limited current (resistors) when the voltage is unknown.

When the voltage is uncertain and the motor has no electrical problems, the following procedures should be followed:

1. Connect it wye and run it without a load, *with limited current.*
2. If the motor starts slowly and draws very few amperes, then the voltage is too low or the delta connection should be used.
3. Connect the motor delta and apply the same voltage (using limited current). The motor will start more quickly and draw more amperes than with the wye connection.
4. Apply full voltage while watching the way it starts, and the no-load amperes.

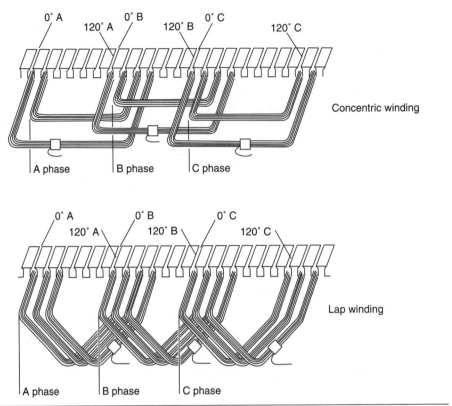

FIGURE 5.55 Two shapes (concentric and lap) used in mush-wound motor windings.

5. If it starts slowly and amperes are low, apply the next-higher voltage, using the same procedure (wye connection first), until the motor starts and runs normally.

6. Load the motor and check the amperes.

7. Assuming the amperes aren't excessively high, use a tachometer to see if the RPM is satisfactory. (Motors above 50 horsepower will have less slip than motors under 50 horsepower.) Divide 7200 by the number of poles to get the motor's synchronous speed (explained in Chapter 3 under "Revolutions per Minute, Poles, and Hertz"). Slip should be less than 1 percent unless the motor is designed for high slip (5 percent slip is maximum for high-slip motors).

8. If slip is excessive with the wye connection, connect the motor delta and run it with limited current and no load.

9. If amperes aren't excessively high, apply full voltage and full load and check the RPM. A motor-connected wye using the delta voltage will

start and run almost normally. With the motor fully loaded, however, the slip will be excessive, the windings will overheat, and the motor will fail in a few hours.

Twelve-Lead Single-Speed Motors

The 12-lead connection (numbered 1 through 12) will have six pairs of leads that light to each other. Figure 5.56 shows the connection. The difference between this connection and the nine-lead connection is that the end of each phase (T10, T11, and T12) is accessible in the motor's connection box.

The 12-lead motor can be connected to four different voltages: wye-delta, one-delta start, two-wye run, and variations of part-winding start methods.

Figure 5.57 is the two-delta connection, which has the lowest voltage rating. Regardless of what the two-delta connection voltage rating is, the voltage rating of the two-wye connection is 1.73 (square root of 3) times that voltage (Fig. 5.58). In this case, if the two-delta connection is rated 220 volts, 220 × 1.73 = 380. The two-wye connection is rated 380 volts.

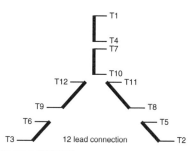

FIGURE 5.56   The 12-lead numbering sequence used in the United States.

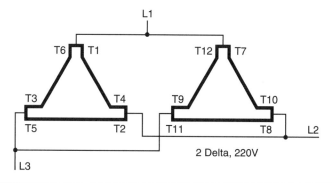

FIGURE 5.57   The lowest voltage connection for the 12-lead motor is two delta.

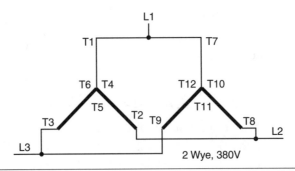

FIGURE 5.58 The voltage rating for the two-wye connection is 1.73 times the two-delta connection.

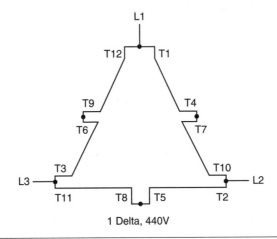

FIGURE 5.59 The voltage rating for the one-delta connection is twice that of the two-delta connection.

If a two-delta (parallel) connection is rated 220 volts, a one-delta (series-connected) is rated 440 volts (Fig. 5.59).

If a two-wye (parallel) connection is rated 380 volts, a one-wye (series-connected) is rated 760 volts (Fig. 5.60).

The 12-lead connection can be used as a part-winding start on two voltages. It can be connected two wye to start and two delta for run (operating on the two-delta voltage). It can also be connected one-wye start and one-delta run (operating on the one-delta voltage).

Another start method (made possible by the 12-lead connection) is one-delta start and two-wye run. This connection will produce 40 percent more starting torque than the wye-start, delta-run connection. (The increased starting torque will require higher inrush amperes.)

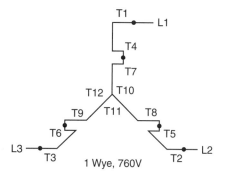

FIGURE 5.60   The voltage rating for the one-wye connection is twice that of the two-wye connection.

The operating voltage must be that of the two-wye connection and may require rewinding the motor for this voltage.

Dual-Voltage Multispeed Motors

The dual-voltage connection is rarely encountered in two-speed motors. As shown in the following schematics, dual-voltage two-speed motors require 18 leads. The schematics in Figs. 5.61, 5.62, and 5.63 (a and b) are constant-torque, constant-horsepower, and variable-torque dual-voltage connections, respectively.

The two-speed dual-voltage-connected winding can't be used with another winding. There are many closed-loop parallel circuits with the two-speed dual-voltage motor connected low voltage (5.63b). All the closed-loop circuits have to be opened when this winding is idle.

European Connections

The International Electrotechnical Commission (IEC) labeling system is used in Europe and Great Britain. Figures 5.64 and 5.65 show this labeling system. Much of the equipment brought into the United States from Europe and Great Britain has motors with this lead numbering system.

Figures 5.66 and 5.67 show the former European connections for the wye-delta and 12-lead motors. Figures 5.68 and 5.69 give the former Great Britain connections for the wye-delta and 12-lead motors.

Most of these motors are designed to operate on 50 Hz. The horsepower ratings are usually given in kilowatts. (Conversion to 60 Hz was explained earlier in this chapter, under "Operation of the Three-Phase Motor.")

Low Voltage

Low speed, 2 delta
L1 – T6, T41, T61
L2 – T2, T52, T62
L3 – T3, T43, T53

T4, T14, T34
T5, T25, T35
T6, T16, T26

High speed, 4 wye
L1 – T6, T16, T26
L2 – T5, T25, T35
L3 – T4, T14, T34

T1, T41, T61	All
T2, T52, T62	tied
T3, T43, T53	together

High Voltage

Low speed, 1 delta
L1 – T1   T14, T41
L2 – T2   T16, T61
L3 – T3   T26, T62
            T25, T52
4,5,6 Open T34, T43
            T35, T53

High speed, 2 wye
L1 – T6
L2 – T5   T1,T2, T3
L3 – T4
T14, T41
T16, T61
T26, T62
T25, T52
T34, T43
T35, T53

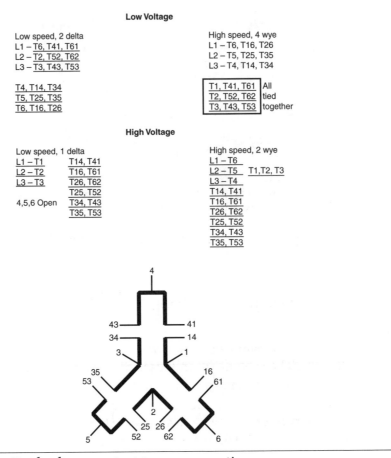

FIGURE 5.61 Dual-voltage constant-torque connection.

European frame sizes are metric. If they are replaced with NEMA frame sizes, an adapter plate can be used.

The European connection system can be replaced with the simpler numbering system found in the United States. The internal connections are identical to those of U.S. motors.

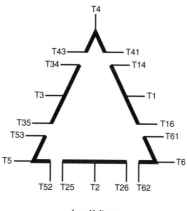

Low Voltage

Low speed, 4 wye
L1 – T1, T41, T61
L2 – T2, T52, T62
L3 – T3, T43, T53

T4, T61, T14	
T6, T52, T26	All tied
T5, T35, T43	together

High speed, 2 delta
L1 – T6, T26, T16
L2 – T4, T14, T34
L3 – T5, T25, T35

T62 & T52 Together
T41 & T61 Together
T53 & T43 Together

T1, T2, & T3 Open and
insulated

High Voltage

Low speed, 2 wye
L1 – T1, L2 – T2, L3 – T3

T16 & T61 Together
T26 & T62 Together
T25 & T52 Together
T35 & T53 Together
T34 & T43 Together
T14 & T41 Together

T4, T5, T6 Together

High speed, 1 delta
L1 – T6, L2 – T4, L3 – T5

T16 & T61 Together
T26 & T62 Together
T25 & T52 Together
T35 & T53 Together
T34 & T43 Together
T14 & T41 Together

T1, T2, T3 Open and
insulated

FIGURE 5.62   Dual-voltage constant-horsepower connection.

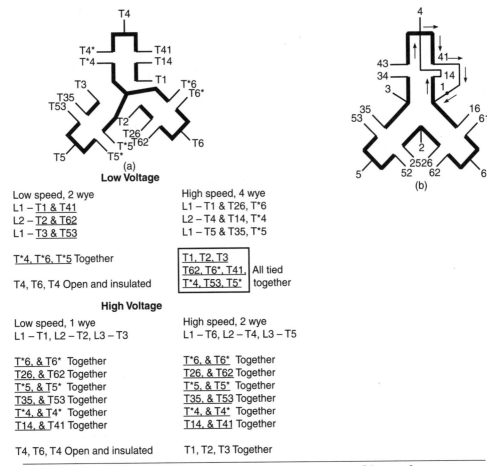

Low Voltage

Low speed, 2 wye
L1 – <u>T1 & T41</u>
L2 – <u>T2 & T62</u>
L1 – <u>T3 & T53</u>

High speed, 4 wye
L1 – T1 & T26, T*6
L2 – T4 & T14, T*4
L1 – T5 & T35, T*5

<u>T*4, T*6, T*5</u> Together

T4, T6, T4 Open and insulated

<u>T1, T2, T3</u>	
<u>T62, T6*, T41,</u>	All tied
<u>T*4, T53, T5*</u>	together

High Voltage

Low speed, 1 wye
L1 – T1, L2 – T2, L3 – T3

High speed, 2 wye
L1 – T6, L2 – T4, L3 – T5

<u>T*6, & T6*</u> Together
<u>T26, & T62</u> Together
<u>T*5, & T5*</u> Together
<u>T35, & T53</u> Together
<u>T*4, & T4*</u> Together
<u>T14, & T41</u> Together

T4, T6, T4 Open and insulated

<u>T*6, & T6*</u> Together
<u>T26, & T62</u> Together
<u>T*5, & T5*</u> Together
<u>T35, & T53</u> Together
<u>T*4, & T4*</u> Together
<u>T14, & T41</u> Together

T1, T2, T3 Together

FIGURE 5.63   (a) Dual-voltage variable-torque connection. (b) Circulating current (arrows) if connection isn't open when used with another winding.

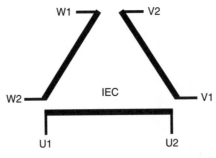

Wye <u>L1-U1</u>, <u>L2-V1</u>, <u>L3-W1</u>, <u>V2&U2&W2</u>
Delta <u>L1-U1&W2</u>, <u>L2-V1&U2</u>, <u>L3-W1&V2</u>

FIGURE 5.64   The latest version of the wye-delta connection, developed by the IEC in Europe and Great Britain.

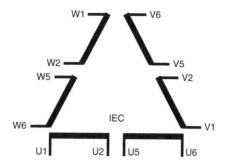

Wye high voltage <u>L1-U1</u>, <u>L2-V1</u>, <u>L3-W1</u>, <u>U2&U5</u>, <u>V2&V5</u>,
<u>W2&W5</u>, <u>U6&V6&W6</u>
Wye low voltage <u>L1-U1&U5</u>, <u>L2-V1&V5</u>, <u>L3-W1&W5</u>
<u>U2&U6&W2&W6&V2&V6</u>
Delta high voltage <u>L1-U1&W6</u>, <u>L2-V1&U6</u>, <u>L3-W1&V6</u>
<u>U2&U5</u>, <u>V2&V5</u>, <u>W2&W5</u>
Delta low voltage <u>L1-U1&U5&W2&W6</u>, <u>L2-V1&V5&U2&U6</u>
<u>L3-W1&W5&V2&V6</u>

FIGURE 5.65   The IEC 12-lead numbering system for Europe and Great Britain.

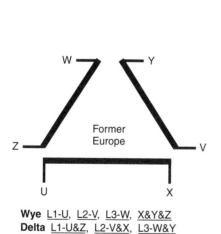

Wye <u>L1-U</u>, <u>L2-V</u>, <u>L3-W</u>, <u>X&Y&Z</u>
Delta <u>L1-U&Z</u>, <u>L2-V&X</u>, <u>L3-W&Y</u>

FIGURE 5.66   The former wye-
delta connection
for Europe.

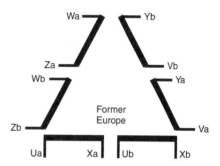

Wye high voltage <u>L1-Ua</u>, <u>L2-Va</u>, <u>L3-Wa</u>, <u>Xa&Ub</u>, <u>Ya&Vb</u>,
<u>Za&Wb</u>, <u>Xb&Yb&Zb</u>
Wye low voltage <u>L1-Ua&Ub</u>, <u>L2-Va&Vb</u>, <u>L3-Wa&Wb</u>
<u>Xa&Xb&Ya&Yb&Za&Zb</u> together
Delta high voltage <u>L1-Ua&Zb</u>, <u>L2-Va&Xb</u>, <u>L3-Wa&Yb</u>
<u>Xa&Ub</u>, <u>Ya&Vb</u>, <u>Za&Wb</u>
Delta low voltage <u>L1-Ua&Ub&Za&Zb</u>, <u>L2-Va&Vb&Xa&Xb</u>
<u>L3-Wa&Wb&Ya&Yb</u>

FIGURE 5.67   The former 12-lead
connection for Europe.

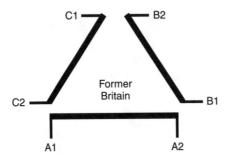

Wye con. <u>L1-A1</u>, <u>L2-B1</u>, <u>L3-C1</u>, <u>A2&B2&C2</u>
Delta con. <u>L1-A1&C2</u>, <u>L2-B1&A2</u>, <u>L3-C1&B2</u>

FIGURE 5.68   The former wye-delta connection for Great Britain.

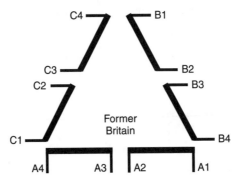

Wye high voltage <u>L1-A4</u>, <u>L2-B4</u>, <u>L3-C4</u>, <u>A2&A3</u>, <u>B2&B3</u>, <u>C2&C3</u>, <u>A1&B1&C1</u>
Wye low voltage <u>L1-A2&A4</u>, <u>L2-B2&B4</u>, <u>L3-C2&C4</u>, <u>A1&A3&B1&B3&C1&C3</u>
Delta high voltage <u>L1-A4&C1</u>, <u>L2-A1&B4</u>, <u>L3-B1&C4</u>, <u>A2&A3</u>, <u>B2&B3</u>, <u>C2&C3</u>
Delta low voltage <u>L1-A2&A4&C1&C3</u>, <u>L2-B2&B4&A1&A3</u>, <u>L3-C2&C4&B1&B3</u>

FIGURE 5.69   The former 12-lead connection for Great Britain.

Chapter 5 Review

1. What qualities make the three-phase induction motor popular for industry (page 201)?

2. The speed of a three-phase induction motor is governed by the number of poles, the Hz of the power source, and the load (page 201). T___ F___

3. The laminations in a 25-Hz motor core are thicker than those in a 60-Hz motor (pages 203–204). T___ F___

4. A motor with a short body is a good choice for vibrating loads (page 204). T___ F___

5. The three phases of a three-phase motor are separated from each other by _____ electrical degrees (page 205).

6. Rotating torque occurs because of time that peak power develops in each phase matches the degrees of physical separation of the phases (pages 206–208). T___ F___

7. Single-phase voltage will not start a three-phase motor (page 209). T___ F___

8. A three-phase motor running on single phase will have about _____ of its rated horsepower (pages 209–210).
   a. $^1/_3$
   b. $^1/_2$
   c. $^3/_4$

9. An induction motor draws high amperes when it starts because of the high current demand of the squirrel cage rotor (page 210). T___ F___

10. High (starting) amperes mean high torque (page 210). T___ F___

11. The code letter is used to determine a motor's full-load amperes (page 213). T___ F___

12. A motor's torque value is constant as the RPM change (pages 213–214). T___ F___

13. A load has torque, RPM, and horsepower requirements, whereas a motor needs only the horsepower rating that the load demands (page 214). T___ F___

14. Nameplate RPM (full-load RPM) is the speed at which a motor has the most torque (page 214). T___ F___

15. The design letter helps fit a motor to its load requirements (page 214).
T___ F___

16. The design A motor is used when nearly constant-load RPM is required (pages 214–215). T___ F___

17. The design B motor is popular because it meets most load requirements, and doesn't require high starting current (page 215).
T___ F___

18. The design C motor has higher torque characteristics than the design B motor (pages 215–216). T___ F___

19. The design D motor draws low starting amperes and has high starting torque, but has low torque at full RPM (page 216). T___ F___

20. The design D motor can handle intermittent overloads that would overheat design B or C motors (page 216). T___ F___

21. All induction motors operate at exactly nameplate RPM (pages 216–217).
T___ F___

22. A wye-connected motor has four circuits, and the delta-connected motor has three circuits (page 217). T___ F___

23. The nine-lead delta connection has one circuit of three leads and three circuits of two leads (pages 217–218). T___ F___

24. A telephone keypad can be used to identify numbers on the 9-, 10-, and 12-lead connections (pages 218–219). T___ F___

25. The 12-lead connection has four voltage connections, and more starting options than the 9-lead connection (pages 224–227). T___ F___

26. The 12-lead connection can be used for part-winding start and other start methods (page 225). T___ F___

27. Multispeed motors all have one winding (page 227). T___ F___

28. Name three types of one-winding two-speed motors (pages 227–228).

29. A two-speed one-winding motor, with no load, draws less current on low speed than it does on high speed (pages 228–230). T___ F___

30. A constant-horsepower motor has high torque on low speed (pages 228–229).
T___ F___

31. A constant-torque motor has more horsepower on _____ speed (page 229).
    a. low
    b. high

32. A variable-torque motor has a horsepower rating similar to a constant-torque motor (pages 229–230). T___ F___

33. Two-speed two-winding motors have no internal connection between the windings (pages 230–231). T___ F___

34. The multimode motors' main feature is to save energy cost during low loading for prolonged periods of time (page 231). T___ F___

35. The multimode motor is used because it has a good power factor, and because its power can be adjusted to different load requirements (page 231). T___ F___

36. The multimode motor is seldom found outside the oil industry, but would work well on an air compressor application (page 231). T___ F___

37. The nine-lead motor that can be confused with the multimode motor when testing for the number of circuits is (page 232)
    a. wye.
    b. delta.

38. The difference between a triple-mode motor and a quadruple-mode motor is the location of the center lead in each phase (pages 233–237). T___ F___

39. The DC field on a synchronous motor is used only for starting the load (page 240). T___ F___

40. The amortisseur winding is a squirrel cage winding (page 240). T___ F___

41. A discharge resistor is used to limit the DC field amperes (page 240). T___ F___

42. High amperes are the main cause of DC field failure (pages 240–241). T___ F___

43. _____ allows the field control relay to energize the DC field (pages 240–241).
    a. Lowered voltage
    b. Low Hz

44. Overexciting the DC field changes the synchronous motor's RPM (pages 241–242). T___ F___

45. The synchronous motor (used to improve an industry's power factor) has to be loaded (pages 243–244). T___ F___

46. The brushless synchronous motor operates the same as the brush type (pages 243–244). T___ F___

47. The synchronous three-phase motor can be converted to an alternator that produces three-phase power (page 244). T___ F___

48. The control that was formerly used to improve power factor on a synchronous motor is used to control the Hz in an alternator (page 244). T___ F___

49. The amortisseur (squirrel cage) winding becomes a damper winding in an alternator (page 244). T___ F___

50. The wound rotor motor is an induction motor (pages 244–245). T___ F___

51. The three-phase winding (and controller) on the rotor of a wound rotor motor controls the motor's torque (pages 245–247). T___ F___

52. Lowering the rotor amps lowers the stator amps (pages 247–248). T___ F___

53. Eliminating resistance in the rotor control too fast stresses the motor's windings (pages 247–248). T___ F___

54. The wound rotor control precisely regulates the motor's speed even if there is no load (page 248). T___ F___

55. The wound rotor motor can be reversed by interchanging its _____ leads (page 248).
    a. rotor
    b. stator

56. A wound rotor motor can be used as a variable-voltage transformer or as a Hz control (page 249). T___ F___

57. A brushless DC motor runs at synchronous speed (pages 249–250). T___ F___

58. Servo motors are a miniature brushless DC motor (pages 251–252). T___ F___

59. An unidentified nine-lead dual-voltage motor may be connected wye, delta, or multimode (page 252). T___ F___

60. An unidentified three-lead three-phase motor should be test run with low voltage and limited current before trying a higher voltage (page 253). T___ F___

61. Six-lead motors are the hardest to identify (pages 253–254). T___ F___

62. Name the three types of one-winding two-speed motors (page 254–256).

63. It isn't advisable to first test run an unidentified motor with full voltage (page 253). T___ F___

64. Higher lead numbers indicate a higher speed winding in a two-winding two-speed motor (page 256). T___ F___

65. The main purpose of the part-winding start is to reduce the starting amps (pages 256–257). T___ F___

66. A two-pole motor may not start using the part-winding start method (page 257). T___ F___

67. The standard concentric-wound motor may not start with half its winding (pages 257–258). T___ F___

68. A wye-delta-connected motor has two circuits with three leads (page 258). T___ F___

69. The wye-delta start method allows more time to start a load without overheating the winding (pages 258–259). T___ F___

70. Once the connection and voltage are identified, the tachometer should be the final test instrument used (at full load) (pages 259–260). T___ F___

71. In the standard 12-lead motor, lead T10 is the end of the A phase, T11 is the end of the C phase, and T12 is the end of the B phase (pages 261–262). T___ F___

72. The 12-lead motor can be used for part-winding start (page 262). T___ F___

73. Dual-voltage two-speed motors need 18 leads (page 263). T___ F___

74. The European connection system can't be converted to the U.S. number system without altering the internal connections (page 263). T___ F___

Chapter 6

Troubleshooting Three-Phase Motors

The Dual-Voltage Nine-Lead Motor's Internal Connections

Costly mistakes can be avoided if the technician understands the two types of connections used in the nine-lead motor. Testing can be done accurately and with confidence if the connections are fully understood.

The dual-voltage nine-lead three-phase motor is internally connected wye or delta. Data found between lead numbers are very different in the two connections. Some winding failures in a wye-connected motor will have a different look from those in a delta-connected motor—even when both have the same basic problem. Test results will also be different.

It's important to identify the cause of burned windings. They will have a different appearance in all these situations: single-phase burnout, overload, unbalanced voltage, and voltage spikes. (Voltage spike damage occurs more often in motors controlled by variable-hertz drives.) These problems are all caused by in-plant faults that require correction. A replacement motor can fail—sometimes immediately—if the in-plant problem isn't corrected.

Nine leads make it possible to operate a three-phase motor on either of two voltages. The two connections have a 2-to-1 voltage difference.

Understanding the Nine-Lead Wye Connection

The nine-lead wye-connected motor consists of three single-phase windings. Each phase has identical turns, coils, and wire size. In addition, they are

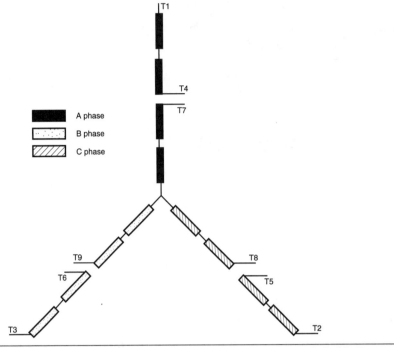

FIGURE 6.1 The four-pole nine-lead motor with two poles in each half of its phase.

each separated into two equal circuits. The end of each phase is connected together, forming a Y connection (Fig. 6.1).

The two circuits of each phase can be externally connected in series or parallel. Series is for high voltage, and parallel is for low voltage. The high-voltage connection (series) is always rated twice as high as the low-voltage connection (parallel).

The connection schematic (Fig. 6.2) can be used as a guide for testing any nine-lead wye-connected motor. Identical length between numbers on a schematic means the data are the same, that is, T1 to T4, T2 to T5, T3 to T6, T7 to T8 and to T9. Identical circuits can be comparison-tested, which is an easy, reliable, and accurate method of checking a winding. Identical faults in each identical circuit (six total) are possible, but extremely remote.

Internally, a winding may have parallel wye circuits (Fig. 6.3). This reduces the winding's wire size. (Smaller wire is easier to form and insert into the slots.) Motors that are 5 horsepower and larger usually have multiple internal wye connections.

If there is more than one internal wye connection, each half circuit will have one or more parallel circuits. An open circuit in a parallel connection

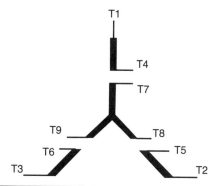

FIGURE 6.2 Using the nine-lead schematic to do the comparison test.

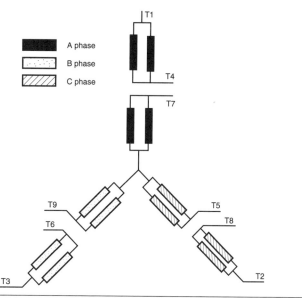

FIGURE 6.3 A four-pole two- and four-wye nine-lead motor.

(multiwye) will show test results different from those of an open circuit in a one-wye connection (Fig. 6.4).

Lead numbers T1 and T4 are the first half of the phase A circuit. Lead T7 starts the remaining half of the A-phase circuit and ends at the internal wye.

Lead numbers T2 and T5 are the first half of the phase C circuit. Lead T8 starts the remaining half of the C-phase circuit and ends at the internal wye.

Lead numbers T3 and T6 are the first half of the phase B circuit. Lead T9 starts the remaining half of the B-phase circuit and ends at the internal wye.

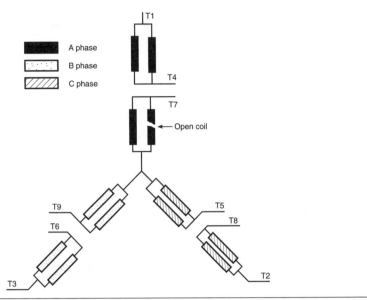

FIGURE 6.4 A four-pole four-wye nine-lead motor with an open circuit.

Leads T1 and T4, T2 and T5, and T3 and T6 have exactly the same data between them. Leads T7, T8, and T9 have twice as many turns (between them) as the three pairs have.

With the low-voltage (parallel) connection, line 1 connects to T1 and T7; line 2 to T2 and T8; and line 3 to T3 and T9. Leads T4, T5, and T6 are connected together and insulated, forming an external wye (Fig. 6.5). Leads T1, T2, and T3 have identical data to leads T7, T8, and T9.

With the high-voltage (series) connection, line 1 connects to T1, line 2 to T2, and line 3 to T3. Lead T4 connects to T7, T5 connects to T8, and T6 connects to T9. All are insulated separately (Fig. 6.6).

The motor winding data are the same from line 1 to line 2, line 2 to line 3, and line 3 to line 1.

Current flow from line to line in a wye connection will pass through two complete phases (Fig. 6.7).

Understanding the Nine-Lead Delta Connection

The nine-lead delta-connected motor consists of three single-phase windings. Each phase has identical turns, coils, and wire size. Each phase end is connected to the start of the next phase (Fig. 6.8).

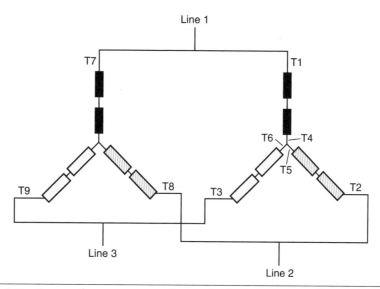

FIGURE 6.5   Low-voltage-connected nine-lead wye-connected motor.

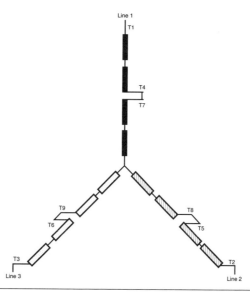

FIGURE 6.6   High-voltage-connected nine-lead wye-connected motor.

Each phase is separated into two equal circuits. The two circuits can be connected in series or parallel. The series connection is high voltage, and the parallel connection is low voltage. The high-voltage (series) connection will always be rated twice the low-voltage (parallel) rating.

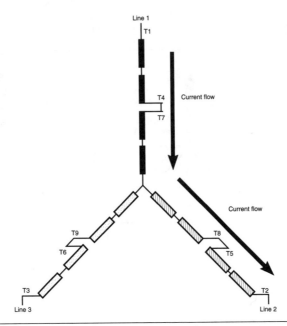

FIGURE 6.7   Current flows through two phases from L1 to L2.

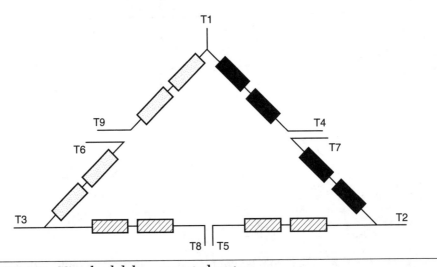

FIGURE 6.8   Nine-lead delta-connected motor.

The connection schematic (Fig. 6.9) can be used as a guide for testing any nine-lead delta-connected motor. Identical length between numbers means the data are identical between them, that is, T1 to T4 and to T9; T2 to T5 and to T7; and T3 to T6 and to T8. Identical circuits can be comparison-

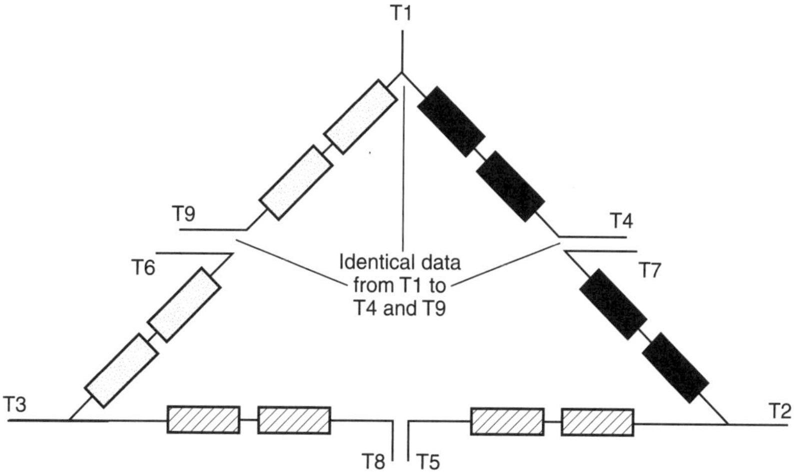

FIGURE 6.9   Using the nine-lead delta schematic to test the motor's circuits.

tested. Identical faults in each phase circuit (six circuits total) are possible, but extremely unlikely.

Smaller Wire Size with Delta and Multicircuit Winding

The delta-connected winding uses smaller wire than the wye connection. Multiple internal circuits also use smaller wire than a single circuit. (Smaller wire is easier to form and insert into the slots.)

The delta connection is usually found in 5-horsepower motors (and larger), but any size three-phase motor can be delta connected. The same is true of multiple-circuit connections. Internally, a winding can have many parallel circuits (Fig. 6.10).

The dual-voltage (nine-lead) delta connection, with more than one internal delta circuit, is shown in Fig. 6.11. Each half circuit (for example, T1 to T4) has two parallel circuits.

The following list refers to the schematic in Fig. 6.12:

- Leads T1 and T4 are the first half of the phase A circuit. Lead T7 starts the remaining half of the A-phase circuit and ends at line 2 (T2).
- Leads T2 and T5 are the first half of the phase C circuit. Lead T8 starts the remaining half of the C-phase circuit and ends at line 3 (T3).

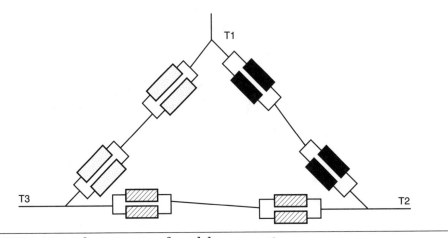

FIGURE 6.10 A four-circuit or four-delta connection.

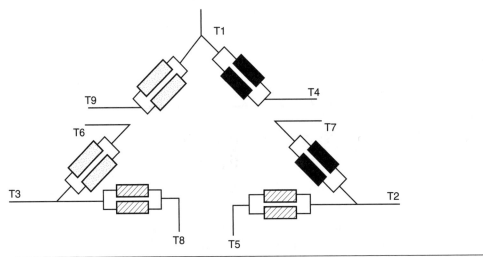

FIGURE 6.11 The nine-lead delta schematic with parallel internal connections.

- Leads T3 and T6 are the first half of the phase B circuit. Lead T9 starts the remaining half of the B-phase circuit and ends at line 1 (T1).
- Data between T1 and T4 are exactly the same as those between T1 and T9.
- Data between T2 and T5 are exactly the same as those between T2 and T7.
- Data between T3 and T6 are exactly the same as those between T3 and T8.

With the low-voltage (parallel) connection (Fig. 6.13), line 1 connects to T1, T7, and T6; line 2 to T2, T8, and T4; and line 3 to T3, T9, and T5.

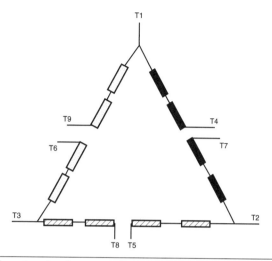

FIGURE 6.12   The nine-lead one- and two-delta connection.

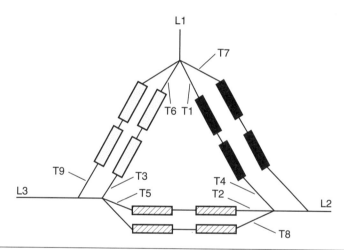

FIGURE 6.13   The nine-lead one- and two-delta connection for low voltage.

In the high-voltage connection (Fig. 6.14), line 1 connects to T1, line 2 to T2, and line 3 to T3. Lead T4 connects to T7; T5 to T8; T6 to T9; the three connections are insulated separately.

The number combination for high voltage is the same for both wye (Fig. 6.15) and delta. The data between line 1 and line 2, line 2 and line 3, and line 3 and line 1 are exactly the same in both high- and low-voltage connections.

Most of the current that flows from line 1 to line 2 (in a delta connection) will pass through one complete phase winding (Fig. 6.16). Some current will flow through the other two phases.

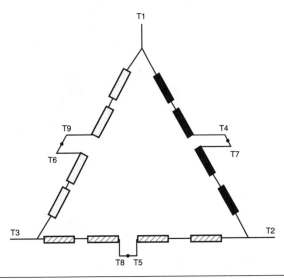

FIGURE 6.14 The nine-lead one- and two-delta connection for high voltage.

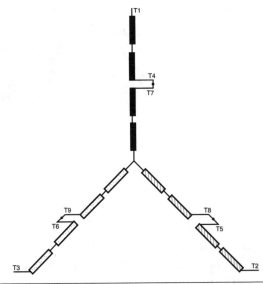

FIGURE 6.15 The nine-lead one- and two-wye connection for high voltage.

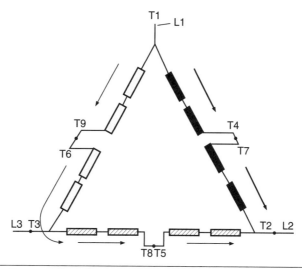

FIGURE 6.16   Current flow from L1 to L2 in the nine-lead delta-connected motor.

Identifying Unmarked Leads in a Nine-Lead Three-Phase Motor

Some methods of identifying unmarked leads require running the motor on part of its winding. Motors with concentric-shaped coils won't start by themselves on part of their winding. The shaft has to be spun to start them, which is dangerous. Voltage readings must be taken from the leads with the motor running. (This is also dangerous.)

The following method is much safer and simpler. (Use the comparison test on all circuits to be sure there is no winding failure.)

Identifying Unmarked Leads in a Wye Connection

The following equipment is needed:

- Test light or ohmmeter
- Low-voltage DC source (6- or 12-volt battery)
- Low-scale DC voltmeter (*must be analog*)
- Numbered lead labels

The motor must be assembled (the rotor is needed to complete the magnetic circuit).

The first step is to separate the circuits by using an ohmmeter or test light. As can be seen in Fig. 6.17, there are three sets of two leads that light together and one set of three leads. Pair off the sets of two. Permanently label the leads in the set of three with numbers T7, T8, and T9 as shown.

Next, locate T1 and T4. Connect one of the DC voltage-source leads to T8, then make and break (flash) the circuit by intermittently touching the other lead to T9. Check across all three (two-lead) sets with the voltmeter while flashing T9. The voltmeter will show a strong deflection on two of the pairs and little or no deflection on one pair (Fig. 6.18). The pair with little

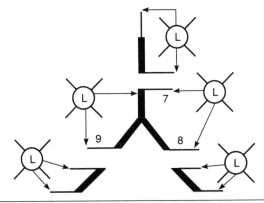

FIGURE 6.17   Identifying the circuits of a nine-lead wye-connected motor with unmarked leads. The set with three leads is permanently labeled 7, 8, and 9.

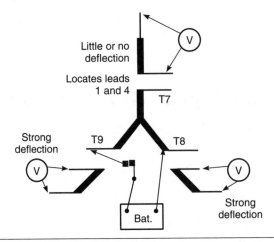

FIGURE 6.18   The battery connection for locating T1 and T4.

or no deflection is T1 and T4—located but not identified—and is part of the T7 phase.

Attach the voltmeter to the located pair. Attach the voltage source to T7 and T8 with the positive lead on T7 (Fig. 6.19). Flash the circuit. If the voltmeter deflects upscale, permanently label the motor lead that is connected to the positive voltmeter probe T1, and its paired lead T4.

With the voltage source still at this location (flashing T7 and T8), check the other two pairs with the voltmeter. The pair with a strong deflection will belong to T8, and the pair with little or no deflection to T9 (Fig. 6.20).

Connect the voltmeter to the pair with a strong deflection (Fig. 6.21). Attach the voltage source to T7 and T8 (with the positive probe on T8). While flashing the circuit (if there is an upscale deflection), permanently

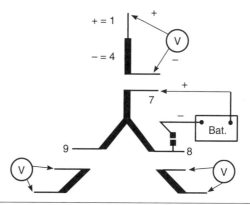

FIGURE 6.19   The connection for identifying T1 and T4.

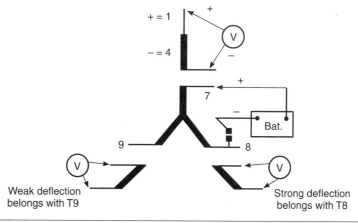

FIGURE 6.20   The battery connection for locating T2 and T5 and also T3 and T6.

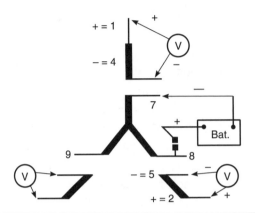

FIGURE 6.21   The connection for identifying T2 and T5.

label the motor lead connected to the positive voltmeter probe T2 and its paired lead T5.

Repeat the preceding procedure—with the positive probe of the voltage source on T9—to identify the last pair (T3 and T6).

In the preceding test, two of the three phases had half of their windings energized, and one complete phase winding was idle. The pair of leads (belonging to the idle phase) had little or no deflection, because its winding wasn't in the same magnetic circuit (angle) as the energized phases. (The idle winding is offset 120 electrical degrees from the others.)

Identifying Unmarked Leads in a Delta Connection

This equipment is needed:

- Test light or ohmmeter
- Low-voltage DC source (6- or 12-volt battery)
- Low-scale DC voltmeter (*must be analog*)
- Numbered lead labels

The motor must be assembled because the rotor is needed to complete the magnetic circuit.

First, separate the circuits. As can be seen in Fig. 6.22, there are three sets of three leads that light to each other.

Next locate the center leads (delta points) of each circuit. (The points of the delta connection—leads T1, T2, and T3—often have larger wire than the

FIGURE 6.22   Identifying the circuits of a nine-lead delta-connected motor with unmarked leads.

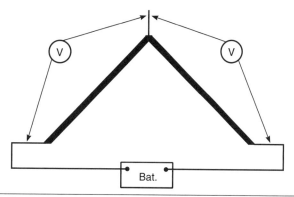

FIGURE 6.23   The connection for identifying the points T1, T2, and T3 of the delta connection.

other six leads.) Attach a DC voltage source to any two leads of any three-lead circuit. There will be a voltage from both voltage source leads—to the point—as shown in Fig. 6.23, if the right leads are selected.

If the point is one of the leads attached to the power source, there will be a voltage read from one source lead to the idle lead and none from the other source lead (Fig. 6.24).

Permanently label the delta points T1, T2, and T3.

Twist the two leads that belong to point T1 together, and apply intermittent DC voltage, as shown in Fig. 6.25. From point T2, one lead (of its three-lead circuit) will show a strong deflection, and the other lead will

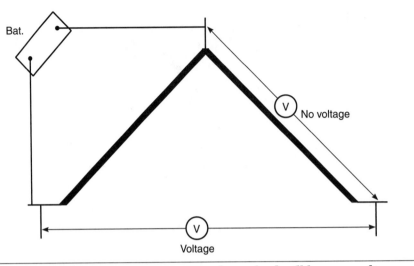

FIGURE 6.24 If the wrong pair is energized, one lead will have no voltage.

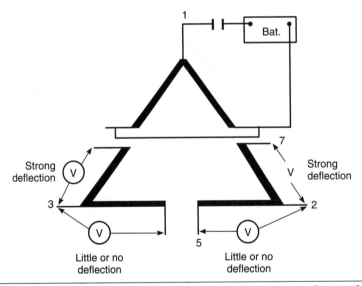

FIGURE 6.25 Power applied to T1 and its internally connected pair identifies T7 and T5.

show little or no deflection. Permanently label the strong deflection lead T7 and the other T5.

From point T3 (Fig. 6.26), one lead (of its three-lead circuit) will show a strong deflection, and the other lead will show little to none. Permanently label the strong deflection lead T6 and the other T8.

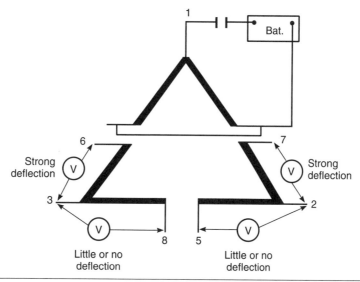

FIGURE 6.26   The voltmeter connection for identifying T8 and T6 and T5 and T7.

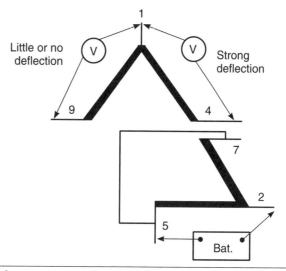

FIGURE 6.27   The connection for identifying T4 and T9.

Connect leads T5 and T7 as shown in Fig. 6.27 and apply intermittent DC.

From point T1, one lead (of its three-lead circuit) will show a strong deflection, and the other lead will show little or none. Permanently label the strong deflection lead T4 and the other T9.

Identifying unmarked leads with the delta connection involves deflection of the voltmeter needle—not polarity as with wye. Half of the two phases that are energized in this procedure are idle. Their windings are cut by the lines of force of the energized half. This produces a strong deflection.

The windings of the idle phase are out of the magnetic circuit and aren't cut by lines of force. Very little or no voltage is transformed into them.

Theory of Unmarked Lead Test

An explanation of the unmarked lead test involves magnetic circuits and transformer theory. Figure 6.28 shows a single-phase magnetic circuit containing four poles.

When half the poles are energized with intermittent DC, two more poles are created in the motor's iron core. The magnetic lines of force of these poles will cut the conductors of the idle poles as shown.

Power is transformed into the idle pole's coils, making it a transformer secondary. An analog DC voltmeter (attached to the idle pole's leads) will show a deflection. (A digital voltmeter reacts too fast and won't show the correct polarity.)

The deflection in the idle half of the winding is strong, because it's in the same magnetic circuit (or angle) as the half that's intermittently energized.

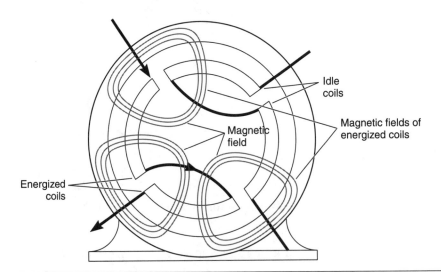

FIGURE 6.28   The magnetic circuit that develops in the stator iron (when half of a phase is energized). The magnetic fields of the energized coils cut the wires of the idle coils, transforming a low voltage into them.

In a three-phase stator, two phases are energized. The third phase is not in the magnetic circuit of the energized phases, and has very little voltage transformed into it.

Typical Winding Problems

It's very important to accurately identify problems that require a motor's removal and replacement. Winding problems that are identified should be documented. A history of the plant's motor problems (on computer software) will point out problem areas that can be improved, or even eliminated.

These winding problems may be found in a three-phase motor:

- Shorted turns
- Ground (winding shorted to frame)
- Phase-to-phase short
- Open winding
- Burned windings from operating on single phase
- Submerged motor
- Assorted rotor problems

These problems require replacing or rewinding the motor.

Shorted Turns

A short is a common winding breakdown, and it requires rewinding or replacing the motor. Shorted turns are caused by nicked coil wire, high-voltage spikes, conductive contaminants, overheated winding, aged insulation, and loose, vibrating coil wires.

As explained under "Inductive Reactance" in Chapter 3, most of the resistance to current flow in an AC motor is furnished by inductive reactance. The resistance of the wire in a complete phase is a very small percentage of the motor's total impedance (resistance plus inductive reactance). *Inductive reactance makes each turn very significant in the motor's ampere demand.* Each turn supplies much more inductive reactance than resistance.

A short forms when one or more turns of a coil are bypassed because of an insulation breakdown between wires. The resistance that the shorted turns develop is eliminated from its phase winding, resulting in increased amperes.

When there are a few shorted turns in one of the three phases, a closed-loop circuit is formed by the turns within the short. As the motor runs, lines of force (from AC current flow) cut the wires in the closed-loop circuit. A high circulating current is transformed into the loop (Fig. 6.29a). Power consumed by the circulating current increases the amperes of the faulty phase, making it easy to identify the problem.

Circulating current in the closed loop often melts the circuit open. When this happens, the circulating current and the turns within the closed loop are eliminated (Fig. 6.29b). Only the resistance of the wire (turns) within the closed loop is now eliminated from the phase winding. Without the ampere demand of the circulating current, the difference lessens between the amperes of the faulty phase and those of the normal phases. A very small difference in resistance is all that is needed to identify the faulty phase. (The rotor should be turned during this test to eliminate its effect.)

Shorted turns in any AC winding are usually visible. They become charred quickly from the high circulating current that is transformed into them (Fig. 6.30).

Ground

When a motor is "grounded," the winding is shorted either to the laminated core or to the motor's frame. The problem is usually found in a slot, where the slot insulation has broken down. Water is the most common cause of a grounded winding. A solid ground requires rewinding or replacing the motor.

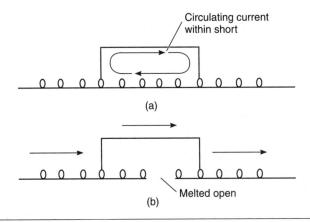

FIGURE 6.29 (a) Transformed current flow in the turns within a shorted coil (closed loop) raises the amperes of the phase. (b) Closed loop melted open.

FIGURE 6.30   A shorted coil. *EASA.*

Some causes of slot insulation breakdowns are overheating, conducting contaminants, lightning, age, pressure of a tight coil fit, hot spots caused by lamination damage (from a previous winding failure), and excessive coil movement. Excessive coil movement is often caused by thermal growth and/or coil twisting torque, brought on by reversing (plugging) or a momentary power interruption.

Phase-to-Phase Short

A phase-to-phase short is caused by insulation breakdown at the coil ends or in the slots. This type of fault requires rewinding or replacing the motor.

Voltage between phases can be very high. When a short occurs, a large amount of winding is bypassed. Both phase windings are usually melted open, so the problem is easily detected.

Among the causes of interphase breakdown are contaminants, tight fit (in the slot), age, mechanical damage, and high-voltage spikes.

Coils that form the poles for each phase are placed on top of each other in all three-phase motors. Figure 6.31 is a concentric-type winding. The coils don't share the slots with other poles in some concentric-type windings.

Figure 6.32 is the lap-winding type. The ends of the coils are nested within each other and have phase insulation between the poles. The coils usually share the slots with other poles. Insulation also separates the coils of each phase in the slots.

Some motors (up to 5 horsepower) are wound with no insulation separating the phases. Phase-to-phase insulation is important because there is a line voltage potential between phases regardless of a motor's horsepower. Figure 6.33 is an end coil phase-to-phase short.

A phase-to-phase short occurs in the slot more often than at the coil ends. When a breakdown occurs in the slot, copper usually melts and fuses to the slot laminations. This copper has to be ground out and removed before the motor is rewound, or it becomes a hot spot and deteriorates the new insulation.

FIGURE 6.31   Windings of a concentric-wound stator. *Electrom Instruments.*

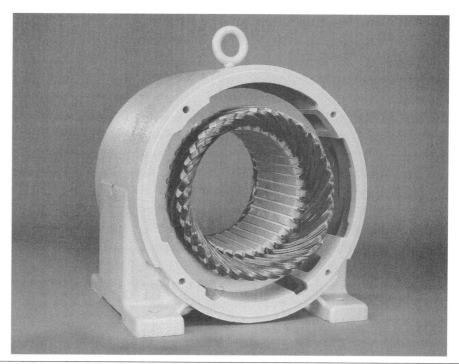

FIGURE 6.32 Windings of a lap-wound stator. *EASA.*

FIGURE 6.33 Winding with a phase-to-phase short. *EASA.*

Open Winding

A common cause of an open winding is undersized lead lugs. (See "Motor Lead Connecting Lugs," later in the chapter.) Charred connections in the motor's connection box are a sure indication of this problem.

Open windings are also caused by shorted turns, phase-to-phase shorts, ground-to-frame shorts, faulty internal coil-to-coil connections, severe overloads, and physically damaged coils. These faults require rewinding or replacing the motor.

An open winding will show several different symptoms (depending on the motor's internal connection). A wye-connected motor with an open winding will test differently from a delta-connected motor.

An open single-circuit winding will be "single-phased." Its power will drop to about half, and the motor won't start.

If the motor's internal connection is multicircuit, it will start but will have reduced power. An open circuit will cause the magnetic circuit to be unbalanced. Under normal load the motor will run more slowly and will overheat. A microhmmeter is used to identify this problem.

A motor with a high number of parallel circuits, that is, four and eight wye, will show less power loss when one circuit is open. Multiparallel circuit connections are used in motors above 5 horsepower.

The windings of a severely overloaded motor (operating on 250 volts) usually become completely charred before an open winding occurs. An overloaded motor operating on 440 volts, however, often will have no sign of burned wires before its windings melt open. In either case, the overload protection isn't working, and the motor should be rewound or replaced.

Burned Windings from Operating on Single Phase

When one line of a three-phase power supply opens, the power becomes single phase. If this happens while the motor is running, its power output is cut approximately in half. It will continue to run, but it can no longer start by itself. Like a single-phase motor without its start winding energized, it has no rotating magnetic field to get it started.

Single-Phase Damage to a Wye-Connected Nine-Lead Motor

Figure 6.34 shows the current path through the wye connection. Two phases of the windings are energized; the third phase has no current flow. If the motor's protection device doesn't function, the two phases that carry

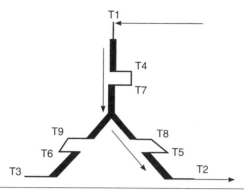

FIGURE 6.34   Schematic showing where the current flows in a wye-connected motor with line 3 open (single phase).

current will overheat and char. The phase without current flow will look normal. Figure 6.35 is a picture of a single-phase-caused burnout in a four-pole winding.

FIGURE 6.35   A wye-connected motor that failed because of a single-phase condition. Line 3 is open. *EASA.*

Single-Phase Damage to a Delta-Connected Nine-Lead Motor

Figure 6.36 shows the current path through the delta connection, with an open phase. The A phase has extremely high current flowing through it. The other two phases have about half as much. The phase with high current will overheat and char if the motor's protection device doesn't disconnect it. The phases that carry less current will look normal. Figure 6.37 is a picture of a single-phase-caused burnout in a four-pole winding.

Submerged Motor

If a three-phase motor has been submerged in water but not energized, there's a good chance it won't need rewinding or replacing. Cleaning and baking the windings may be all that's needed.

The motor should be disassembled as soon as possible. If the motor has ball bearings, they should be replaced. If it has sleeve bearings, the oil wicking material will pit or rust the shaft area located in the bearing window. Replace the oil wick material immediately. If the motor has an oil reservoir and oil ring, the reservoir should be thoroughly cleaned.

The windings should be first tested with an ohmmeter. (A wet winding should never be subjected to a test voltage that could arc through the wet slot insulation.) The baking temperature shouldn't exceed 200°F. The ohmmeter test should read infinity after baking. After the windings have been cleaned, dried, and tested, they will need a coat of air-drying varnish.

When water soaks the slot insulation, the copper windings and the core become a form of battery. A small voltage can be read (with a millivoltmeter)

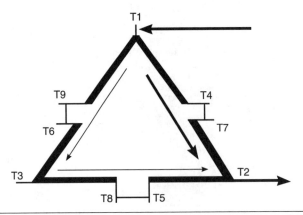

FIGURE 6.36 Schematic showing where the current flows in a delta-connected motor with line 3 open (single phase).

FIGURE 6.37   A delta-connected motor that failed because of a single-phase condition. Line 3 is open. *EASA.*

between the winding and the frame when the slot insulation is wet. A zero reading indicates the motor has been baked long enough.

A megohmmeter, hi-pot, or surge tester can be used when an ohmmeter test shows infinity.

Assorted Rotor Problems

This is a review of the rotor problems found in Chapter 3, with more detailed information:

- Open rotor bars
- Open end rings
- Misaligned rotor/stator iron
- Rotor dragging on the stator
- Rotor loose on shaft

Open Rotor Bars

Open rotor bars or end rings usually necessitate replacing the motor. They can be repaired, recast, or rebarred (if it's economical). (It's important that any replaced metal be the same as the original.)

Open rotor bars are caused by overload burnout, arcing in the slot from a shorted winding, loose bar vibration, thermal growth stress (from starting), flaws in bar material (casting flaw), and poor connections with end rings.

Open rotor bars cause loss of power. If too many rotor bars are open, a loaded motor will draw amperes high enough to open its protection device. With no load, the amperes will be very low. Slow starting and lower-than-rated RPM are a sign of broken rotor bars.

Open End Rings

Open end rings cause uneven torque and some power loss. A ring with one open spot will soon develop more open spots. Each time the open spot crosses a 90° spot between poles, the current will double in the ring area between the next two poles (Fig. 6.38).

Causes of open end rings and/or cracked end rings include flawed casting; motor burned out from overload; motor redesigned for a higher speed (without increased size of end ring); ring material drilled away for balancing; thermal growth stress; and mechanical damage.

A bubble or void in an end ring can cause an electrical vibration. (This type of vibration can't be corrected by balancing.) It can be detected by cutting the power and allowing the motor to coast. Electrically caused vibrations will always cease as soon as the power is shut off.

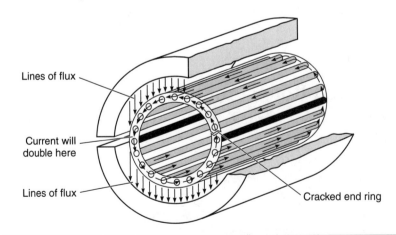

FIGURE 6.38   Current flow in a squirrel cage rotor with an open end ring.

Misaligned Rotor/Stator Iron

A motor with a misaligned rotor will draw high amperes and will lose power. The magnetic path becomes distorted, causing the magnetizing amperes to increase. The stator windings will char and resemble an overload burnout.

Possible causes of a misaligned rotor include:

- Wrong bearing shim placement
- Bearings not installed correctly on the shaft (extended race on wrong side)
- Wrong bearing width
- Captive bearing not held as originally placed
- End bells interchanged
- Stator core shifted on its shell
- A rotor shifted on its shaft
- A rotor replaced with a shorter rotor. A rotor with the same diameter but longer than the original will work, but some efficiency is lost.

Rotor Dragging on the Stator

If the rotor drags on the stator and the bearings aren't worn, it's common practice to "skim" the rotor on a lathe. The process increases the air gap, which increases the no-load amperes. The increased amperes are similar to a misaligned rotor and stator iron. The magnetic circuit is degraded, so it takes more amperes to magnetize the motor's iron. The motor will run hotter than normal, because the motor is drawing more magnetizing amperes. If the load is at maximum or there are any adverse conditions (such as low voltage or frequent starts), the motor should be replaced. (There will be some permanent power and efficiency loss.)

Rotor Loose on Shaft

A loose rotor on a shaft makes a rumbling or vibrating sound. The sound will cease after the power is turned off (while the motor is coasting). If the motor has operated this way for very long, a red dust will form between the shaft and rotor iron. This dust is oxidized iron, caused by the rubbing action between the shaft and the rotor iron. (The same thing happens when a pulley or bearing is loose on a shaft.)

The decision to repair a loose rotor depends on the price of a replacement motor and the importance of the motor in the plant operation. There are options for this problem: The rotor (and shaft) can be replaced;

the rotor can be bored out and a new shaft fitted to it; or a thin metal wedge can be driven between the shaft and rotor to secure it.

Wedging the rotor may offset it enough to make it drag on the stator. It would then be necessary to skim the rotor—on a lathe—to keep it from dragging. The rotor should be bored and fitted with a new shaft. In most cases, it is more economical to replace the motor.

When Motors Overheat

Overheating is a major cause of insulation and bearing failure. Causes of overheating include the following:

- Line
- Operator
- Control
- Motor fault
- Location
- Maintenance

Line-Caused Overheating

There are four causes of line-caused overheating: overvoltage, low voltage, unbalanced voltage, and a high-resistance connection.

Overvoltage

Overvoltage can cause a normally loaded motor to overheat. NEMA standard specifications allow motors a voltage deviation of ±10 percent from their nameplate voltage rating.

All motors will start their load faster when voltage is high. T-frame motors, however, overheat (on overvoltage) more than the older U-frame or high-efficiency motors (even with less than 10 percent overvoltage). T-frame motors have smaller copper wire in their windings and significantly less iron in their core. (They depend on a high volume of air to keep them cool.)

NEMA approved the T-frame motor's design in 1967. This downsized motor was made possible by the development of insulation that could tolerate higher temperature.

A big improvement in the T-frame design is the use of ball bearings instead of sleeve bearings. Ball bearings make it possible to have a closer fit between the rotor and the core, thus reducing the air gap. (Air gap reduces a motor's efficiency more than other forms of loss.) The induction motor and the transformer have very comparable characteristics—both transform power from one winding to another. Unlike a motor, a transformer has no air gap. Therefore, some transformers have efficiency ratings above 99 percent.

Earlier T-frame motors had higher efficiency than U-frame motors, as less iron was used in the core and rotor. (Magnetizing more iron than necessary lowers the efficiency.) Design changes that improved the T-frame motor's efficiency included getting rid of excess iron and reducing the air gap. However, because of the reduced overall size, there was a loss of area for heat dissipation.

Some motor manufacturers reduced material used in the electrical and mechanical components too much. (One manufacturer made fractional-horsepower motors that were too fragile to ship. The stator core was mounted in a very thin steel shell. Several dents were punched in the shell to hold it in place. The core was easily dislodged with only moderately rough handling.)

Manufacturers increased the size of the motor's cooling fan and installed metal baffles to direct the air on the windings. Air volume was increased because of the extra heat created by undersized wire. (Power used to move the extra air reduces the motor's efficiency.)

In one case, there were two motors (made by the same manufacturer) that had different horsepower ratings. They had identical core and rotor dimensions and identical electrical data. One motor was rated $1/3$ horsepower and the other $1/2$ horsepower. The $1/3$-horsepower motor had a totally enclosed frame, and the $1/2$-horsepower motor had an open frame. The nameplate RPM of the $1/2$-horsepower motor was lower than that of the $1/3$-horsepower motor. By loading the motor down a few RPM, it produced $1/2$ horsepower. Air flowing through the open end brackets increased the amount of cooling capability enough to compensate for the increased amperes and heat.

Low Voltage

Low voltage affects all motors in the same way—they lose power and may take too long to start a load. Prolonged starts cause excess heat in the stator winding (at the start of its duty cycle). Another concern with low voltage is the heat that develops from frequent starting.

Under maximum load, low voltage causes the motor to run more slowly than the nameplate RPM. Technically, it's overloaded. It will run hot and will soon fail.

If the service wire to a motor is too small, it causes low voltage (especially when starting). Some symptoms are frequently blown fuses or tripped protection devices. Voltage normally drops briefly from the motor's high starting current, but the voltage should come back up to nearly its full value as soon the motor reaches full speed. Although NEMA standards allow ± 10 percent of nameplate voltage, the voltage should not drop more than 1 percent. (The NEMA standard pertains to the motor's supply voltage, not the voltage drop.) High-efficiency motors draw higher starting current than standard motors. This may require an increase in the wire size of the feeder line.

Motors with a maximum load are the first to react to low voltage. A prime example was a motor that tripped its overload protector at a certain time of the day. The cause was traced to a high-ampere load coming online (elsewhere in the plant) at that time, lowering the voltage for the entire plant.

Power lines that are loaded nearly to their capacity can have intermittent low voltage. A brief high-load demand from another location will cause a voltage drop.

Power companies put capacitors across the lines to correct the low power factor. Inductance develops in power lines over a long distance. Lines are crossed at regular intervals to cancel the inductance. Over many miles, however, inductance still accumulates, requiring capacitors to do the final power factor adjustment. Low voltage will develop if the capacitors fail. A voltage recorder should be used if incoming power problems are suspected. The recorder should be on long enough to include every load variation the location has.

Unbalanced Voltage

Unbalanced voltage has many sources:

- Failed power-factor-correcting capacitors can affect the voltage of an entire building.
- A high-resistance connection is a frequent cause of an unbalance. This problem can be easily spotted with infrared instruments.
- Unbalanced single-phase loading (such as lighting and single-phase motors) is a frequent cause of unbalanced voltage. Lighting load can

be balanced at the distribution panel. (A single-phase air conditioner is an example of a motor load that is hard to balance, due to the air conditioner's intermittent operation.)

Unbalanced voltage will cause a motor to overheat even if it is lightly loaded. Motors are designed to have exactly the same number of turns in each phase. Unbalanced voltage is the same as taking turns from one phase and putting them into another. The result is unbalanced and conflicting magnetism. The distorted location and uneven magnitude of the poles (transformed into the rotor) create bucking magnetic flux and harmful harmonics.

Harmonics are a voltage within a voltage (Fig. 6.39). When the harmonic voltage is on the opposite side of the zero line (from the supply voltage), its value subtracts from the line voltage. This reduces the line voltage value but weakens the motor.

A winding failure caused by unbalanced voltage will have a definite appearance (as shown in Fig. 7.8 in the next chapter). In a wye-connected

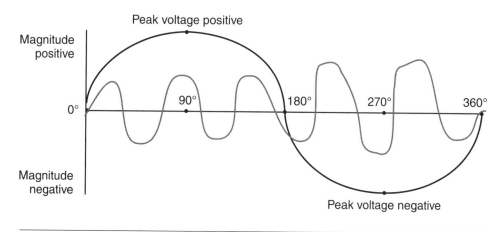

FIGURE 6.39   A harmonic voltage sine wave within a line voltage sine wave. *EASA.*

winding, one phase is charred, one is discolored (from excess heat), and the third phase isn't affected. A delta-connected winding will look similar to the wye-connected winding—two phases are overheated and one phase is unaffected.

Another cause of unbalanced voltage-related heating is worn contacts (in a motor starter) that don't make a solid connection.

High-Resistance Connection

A high-resistance connection is one of the more frequent causes of electric motor failure. Because of a motor's high starting current, its ampere load is very different from most electrical loads. Every connection in the motor's supply circuit is subjected to a brief high-ampere stress when it starts. These connections should be checked regularly (more often if the motor starts frequently).

The National Electric Code specifies wire size for a motor's supply line according to the motor's nameplate amperes. Most motors have inrush current many times their nameplate value (see "Code Letter" in Chapter 5). (Inrush current of high-efficiency motors is higher than that of standard motors.) Increasing the supply line wire size (one size above the code requirement) is a one-time cost that decreases the chance of motor failure.

Connecting Terminals   The connecting terminals in a disconnect switch and/or a motor control are common spots for high-resistance connections. Heat expands the copper wire faster than it expands the control's terminal. The terminal doesn't give, so the wire mashes and is distorted. This has an accumulating effect, loosening the connection slightly each time the motor starts. Eventually a high-resistance connection forms.

Disconnect and control connections should be routinely checked and tightened. A small infrared gun can be used to check connections. All connections should be the same temperature.

Aluminum wire has a notorious reputation for developing loose connections. Its use is not recommended for an electric motor load.

Motor Lead Connecting Lugs   Motor lead lugs should be thick enough (throughout the connection) to represent the circular mil area (size) of the motor's lead wire. If any part of the lug is too small, it becomes a resistor in series with the motor, and current will be restricted when the motor needs it the most—to start the load.

Figure 6.40 shows lugs that aren't made for electric motors. Lug (a) is a piece of copper tubing, which has been partially flattened and has a hole punched in it for the connecting bolt. Its ferrule will hold wire that has a much greater circular mil area than that of the bolted part of the lug. Lug (b) is clearly not a motor duty lug.

Motor Lead Wire Size   Some motor manufacturers size the lead wire according to the motor's nameplate amperes. The wire has high-temperature insulation (usually white) and coarse, tinned strands. When installing lugs on this wire, be sure to include all the strands. One strand represents a high percentage of the wire's (inadequate) total circular mil area.

When an electric motor repair center installs lead wire, the wire is equal to or larger than the total circular mil area of all the internal coil wires that are attached to it. Internal coil connections are twisted together and silver-soldered. The soldered part of the connection joins more circular mil area than that of the coil wires. (This is the last place to get hot from high current.)

Operator-Caused Overheating

Operators often cause motors to overheat. If paid by the unit, operators often overload the machine. The mining industry is a case in point.

An overloaded induction motor can be quickly checked with a tachometer. Compare the tachometer reading with the motor's nameplate.

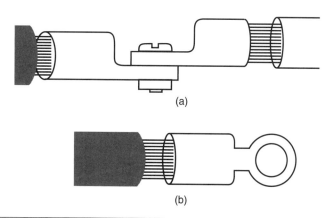

(a)

(b)

FIGURE 6.40   Two lead lugs that can cause motor failure.

Just a few RPM slower than the nameplate RPM will cause overheating. The design A (or a high-efficiency) motor will run hotter than a standard induction motor under this condition. This can be seen in the torque curve that the design A motor develops near its full-load speed (Fig. 6.41).

Frequent starting and stopping often causes excessive heating in motors above 25 horsepower. The physical mass of a large motor doesn't dissipate heat easily.

Most motors will cool better while running with no load than when shut off. If a motor is required to start often, forced ventilation or the use of a clutch should be considered if motor failure is frequent.

An abrupt speed change is harmful to all motors, but especially to brush-type (DC and wound-rotor three-phase) motors. The design D induction motor should be considered if its torque curve is compatible with the load requirement.

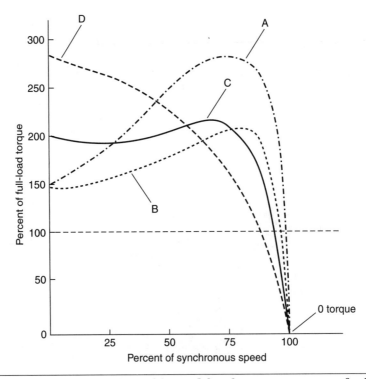

FIGURE 6.41 The design letter table used for choosing a motor to fit the load's torque demands.

Control-Caused Overheating

Duty-cycle loading requirements (that cause overheating) can be relieved by forced ventilation. Some conditions helped by forced ventilation are overheating from prolonged slow speed at maximum torque, frequent start-stop cycle, and a duty cycle that includes overloading for a short time before shutdown. Operating a motor for a time with no load will cool the winding and core uniformly. Running with no load on reduced voltage is an even better way to cool a motor.

The part-winding start method (of starting a motor) isn't recommended for frequent starting. Part of the winding (one-half or two-thirds) starts the load and will heat excessively when started frequently. Switching from part winding to full winding too slowly (over 3 seconds) will also overheat the motor.

If the motor has concentric-wound coils, it won't start on part of its winding. Figure 6.42 compares the concentric- and lap-shaped coils. This motor needs a special internal connection before it can be used on a part-winding start control. Many motors (up to 300 horsepower) use

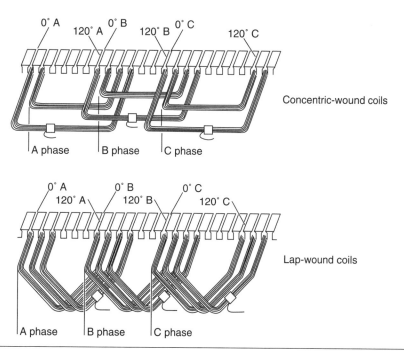

FIGURE 6.42   A motor with concentric-shaped coils won't start using the part-winding start method. However, it will start with lap-shaped coils.

the concentric type of winding. When an old part-winding start motor is replaced, the new motor must be designed for this starting method.

A poorly designed variable-hertz control can cause a motor to overheat. (Voltage should change with the hertz change.) Not all motors are suited for variable-hertz drives. In many cases, older T-frame motors have not worked as well as high-efficiency motors or U-frame motors on variable hertz.

If a motor is labeled *inverter duty* (variable-hertz duty), it doesn't mean the motor is designed for slow speed at maximum torque or any other extreme deviation from the motor's speed. Inverter duty means the motor's coil wire has special insulation. This insulation *postpones* breakdown from voltage spikes caused by abrupt voltage and current changes. It does not *prevent* breakdown. Reactors, filters, or surge capacitors should be installed to reduce this problem.

Heat Developing from Reduced-Power Starting

The amount of internal heat a motor accumulates in across-the-line starting compares to that produced in the wye-delta or reduced-voltage starting methods. Any reduced-power starting method that applies all the motor's winding is better for the motor than a part-winding start method.

Reducing the motor's starting amperes lessens the amount of voltage drop caused by across-the-line starting. In some cases, reduced-power starting is done to protect a load that is sensitive to quick starts.

If a load that requires frequent starting causes the motor to heat excessively, forced ventilation will lengthen a motor's insulation life. Ventilation should continue for a time after the motor is offline. (The amount of time required depends on the motor's size, etc.)

For maximum motor life, it's best to start a large motor with no load. (A clutch can be used to apply the load after the motor has reached full speed.)

Motor Fault Overheating

Shorted Turns

Shorted turns in a winding will cause overheating. When just a few coil turns are shorted, they form a closed loop. A circulating current is transformed into the loop. The current is usually high enough to melt a wire, opening the closed loop.

Before this happens, a pole forms from the circulating current in the shorted turns. This pole doesn't conform to the surrounding poles. Its magnetism causes a ringing sound, unique to motors with shorted turns.

Shorted turns may not slow the motor at all. A line-to-line ampere comparison will show a substantial difference between motor leads, and will identify the problem as shorted turns.

Ground in Winding

If a winding is grounded (shorted) to the stator or frame, it will get hot (from the increased ampere flow). This condition usually causes a fuse or breaker to open. If there are enough coils in the motor's circuitry between the line and the ground, the motor will still run, but with increased (and unbalanced) amperes.

Worn Bearings and Uneven Air Gap

Worn sleeve bearings cause overheating. A three-phase motor's torque is so smooth that it may be necessary to move the shaft to detect a worn sleeve bearing. It's common for a rotor to drag on the stator before the problem is detected.

Uneven air gap from worn sleeve bearings will cause internal heating in some motors. Motors that use an internal circuit connection (that balances the current path through the poles) are less affected by uneven air gap.

Although uneven air gap should be avoided, it won't cause immediate damage to a motor's winding. In the past, most three-phase motors had sleeve bearings. Many ran for years with worn bearings and an uneven air gap—with no electrical problems.

Wrong Service Factor

The *service factor* (found on the motor's nameplate) is the amount of overload a motor can handle (without overheating) for a limited time. The service factor number is a multiplier. The multiplier number times the motor's nameplate amperes is the amount of overload a motor can handle. Service factor numbers are 1, 1.15, and 1.2.

A high service factor usually indicates a well-designed motor. A totally enclosed motor will have a service factor of 1, meaning it can't be loaded higher than nameplate amperes.

Many air compressors are deliberately designed to use the service factor. (They are in a highly competitive market.)

In one case, an air compressor had a 75-horsepower motor with a 1.2 service factor. The compressor was designed to use the full 1.2 service factor value. (When a compressor operates normally, it will cycle for a time unloaded so the motor has time to cool down.) Unfortunately, this motor never ran unloaded because the air volume demand was so high. The motor failed after a few months.

Connected for the Wrong Voltage

If a dual-voltage motor connected for high voltage is connected to low voltage, it will produce only one-fourth of its rated horsepower. It will start much more slowly than normal. Some loads allow this motor to reach nearly normal speed. If an induction motor's RPM is below its nameplate rating, the high slip will cause it to overheat.

If a dual-voltage motor connected for low voltage is connected to high voltage, the results with any type of load are immediate. The motor develops many times its normal starting torque, and it draws so many amperes that its winding is destroyed in a matter of seconds. NEMA standards allow ±10 percent of nameplate voltage. If a fully loaded motor, rated for 220 volts, is connected to 250 volts, it will run hotter than normal (a 12 percent difference). Motors with frequent start cycles will have extreme overheating problems when voltage is this high.

A motor rated for 208 volts but connected to 250 volts will overheat without a load. (The connection is 20 percent over its rated voltage.)

A motor rated for 250 volts but connected to 208 volts can't pull its rated load. It may not start a load requiring high breakaway torque. The motor will work if the load is reduced. (A tachometer should be used to make sure the RPM isn't below the nameplate rating.) Any departure from rated voltage greater than ±10 percent will result in extra heat.

Wrong Hertz

Motors designed for 50 Hz power most of the machinery manufactured in Europe. Problems can occur when this machinery is used in the United States on 60 Hz. A four-pole 50-Hz motor runs 300 RPM faster on 60 Hz. The motor will be overloaded if its load is air or liquid. Conveyer belts and augers will also overload this motor. (Changing the pulley dimension ratio solves the problem for some applications.) Direct-driven loads require major redesigning or replacement.

In one case, the power for an entire facility was converted to 50 Hz, because so much of the equipment used 50 Hz. When failed motors were replaced with 60-Hz motors, they ran hotter than normal on 50 Hz. (The 60-Hz motors have fewer turns per pole than 50-Hz motors.) If the supply voltage is lowered for the 60-Hz motors, they won't run as hot, but power output is lessened.

Internal Motor Problems

Internal motor problems can cause overheating. The problems that follow were covered earlier, under "Assorted Rotor Problems," but are reviewed here briefly.

Rotor/Stator Alignment   If the rotor and stator iron aren't aligned properly, the result is high amperes (loaded or no load) and loss of power. This problem can't be detected with an ohmmeter, or limited current and turning of the shaft.

Open Rotor Bars   Open rotor bars cause power loss. With a normal load, the rotor will run more slowly than the nameplate RPM, resulting in high amperes in both the rotor and the stator windings. (Too much slip increases rotor hertz, which causes higher amperes.)

Cracked End Ring   Cracked end rings cause uneven torque and loss of power. The result is similar to that of open rotor bars.

Air Gap Too Large   If the rotor becomes "out of round," it may drag on the stator core. This condition is corrected by skimming some of the iron off the rotor with a lathe.

Skimming the rotor increases the air gap. Air gap should be kept at a minimum because it's a break in the magnetic circuit. A large air gap creates a large increase in the motor's magnetizing amperes. The motor will run hotter, and there will be a slight power and efficiency loss.

Whether to skim the rotor should be decided on a case-by-case basis. If the motor has more power than needed or its duty (frequent starts, etc.) doesn't cause above-normal heating, skimming the rotor does no harm.

A large air gap can be similar to a misaligned rotor. (It can't be detected with an ohmmeter, or limited current and turning of the shaft.)

Location-Caused Overheating

In a hot location, the surrounding air increases a motor's normal running temperature. If the motor is loaded near its capacity, it will overheat. Even forced ventilation may not help.

If a motor is located too close to a wall or if anything impedes its intake or exhaust airflow, it can overheat (depending on ambient temperature). The recommended clearance is 18 inches on all sides.

Intense sunlight can cause a maximum-loaded motor (or a motor that starts frequently) to overheat.

An enclosure added for protection from the weather can be harmful to a motor, because air movement is restricted. In addition, the enclosure may trap the motor's hot air exhaust. An exhaust fan is often added to remove the hot air from the enclosure. However, forced air movement shouldn't affect the motor's normal air movement.

Maintenance-Caused Overheating

Belts Too Tight

Extra stress is put on bearings if the belts are too tight. If the motor has a long stator, the shaft may bow and misalign the bearing's inner race. There have been cases where the rotor dragged on the stator core (because of overly tightened belts).

Pulley Problems

A replacement pulley that is slightly larger than the original can overload a motor if the load increases with a speed increase.

A worn pulley may make it necessary to tighten the v-belts excessively (to keep them from slipping). (The motor drive pulley is always the one that wears first.)

Chapter 6 Review

1. The dual-voltage nine-lead three-phase motor will be connected (internally) either wye or delta (page 275). T___ F___

2. In-plant power problems will cause immediate motor failures (page 275). T___ F___

3. Name two in-plant problems that will cause the immediate failure of a three-phase motor (page 275).

4. The ends of each phase are joined internally in a wye-connected motor. (pages 275–276) T___ F___

5. The delta connection has each phase end connected to the start of the next phase (pages 278–281). T___ F___

6. The delta connection and multicircuit design allow the use of smaller wire in the winding coils (pages 281–283). T___ F___

7. Why should the motor be assembled when identifying unmarked leads (pages 285–286)?

8. Why is there no voltage detected in the idle phase during the unmarked lead test (pages 285–288)?

9. The delta connection will have three sets of three leads that light to each other (page 288). T___ F___

10. When testing for unmarked leads, an analog DC voltmeter shows a deflection, and works better than a digital meter (pages 292–293). T___ F___

11. The largest amount of resistance to current flow in a three-phase winding is furnished by inductive reactance (pages 293–294). T___ F___

12. Circulating current within a short can char and melt the turns within the short (page 294). T___ F___

13. A phase-to-phase short is usually obvious, and requires replacing or rewinding the motor (pages 295–296). T___ F___

14. Name three causes of an open winding (page 298).

15. A three-phase motor running on single phase will have (page 298)
    a. full power.
    b. $^2/_3$ power.
    c. $^1/_2$ power.

16. A motor that failed from a single-phase line condition—with one phase burned and two phases near normal—is connected (page 300)
    a. wye.
    b. delta.

17. The first test (from winding to frame) on motors that have been subjected to water should be done with an ohmmeter (page 300).
    T___ F___

18. All motors that have been submerged in water must be rewound (pages 300–301). T___ F___

19. An ohmmeter will always detect a grounded motor winding by testing at the motor's disconnect or control (page 301). T___ F___

20. All rotor problems cause instant motor failure (page 302). T___ F___

21. A sign of misaligned rotor/stator iron is high no-load amperes (page 303).
    T___ F___

22. What is an easy way to detect electrically caused vibration (pages 303–304)?

23. An out-of-round rotor that has been "skimmed" with a lathe has no effect on a motor's performance (page 303). T___ F___

24. A rotor that is loose on its shaft should be replaced (pages 303–304).
    T___ F___

25. Overvoltage will affect some motors more than others (pages 304–305).
    T___ F___

26. What is the allowable NEMA standard deviation (pages 304–305)?

27. Most of a motor's efficiency loss is in the air gap (page 305). T___ F___

28. Within the allowable 10 percent voltage variation, low voltage overheats a (fully loaded) motor more than high voltage does (pages 305–306).
    T___ F___

29. Operating on low voltage is similar to overloading a normally loaded motor (page 306). T___ F___

30. Unbalanced voltage is always caused by a faulty transformer, and is a problem the power supplier has to fix (pages 306–308). T___ F___

31. Unbalanced voltage will cause immediate winding failure in a three-phase motor (page 307). T___ F___

32. Unbalanced voltage affects only wye-connected motors (page 308).
    T___ F___

33. High-resistance connections occur more often in a motor supply line than in that of a lighting circuit (page 308). T___ F___

34. Frequent motor starting increases terminal connection problems (page 308). T___ F___

35. A motor that has oversized supply lines can't be protected from overloads (page 308). T___ F___

36. Tightening the terminals in a motor's disconnect switch should be part of the scheduled motor maintenance (page 308). T___ F___

37. Motor lead lugs and lead wire can be undersized because they carry current such a small distance (pages 308–309). T___ F___

38. Undersized lead lugs will eventually overheat and char, causing an open or high-resistance circuit (page 309). T___ F___

39. A tachometer can quickly determine if a motor is overloaded (pages 309–310). T___ F___

40. If a motor overheats because of its stop/start cycle, but not its load, it's better to install a larger motor than to use forced ventilation (page 310). T___ F___

41. Running a motor with no load is the best way to uniformly cool down a motor (page 311). T___ F___

42. Across-the-line starting and reduced-power starting produce approximately the same amount of internal heat (page 312). T___ F___

43. A motor _____ run with shorted turns in its winding (pages 312–313).
    a. may
    b. won't

44. Shorted turns can sometimes be detected by a ringing sound (pages 312–313). T___ F___

45. An uneven air gap will destroy a winding immediately (page 313). T___ F___

46. A motor with a high service factor can run continuously at this value (page 313). T___ F___

47. If a motor connected for high voltage is run on low voltage, it will produce only half its rated horsepower (page 314). T___ F___

48. A motor rated for 250 volts can operate on 208 volts if its full-load speed is close to its nameplate RPM (page 314). T___ F___

49. A four-pole motor designed for 50 Hz—operating on 60 Hz—will run 300 RPM (pages 314–315)
    a. faster.
    b. slower.

50. A 50-Hz motor has _____ on 60 Hz (pages 314–315).
    a. more power
    b. less power

51. A 60-Hz motor has _____ on 50 Hz (page 315).
    a. more power
    b. less power

52. All rotor problems cause a fully loaded motor to lose power and run hot (page 315). T___ F___

53. Most large motors need 18 inches of clearance, for unrestricted ventilation (page 316). T___ F___

Chapter 7

Troubleshooting Less Common Motors

Troubleshooting Logic Using Voltage Rules

Many electrical problems can be explained using the logic in these three rules:

1. When a conductor cuts or is cut by magnetic lines of force, a voltage is created in the conductor.
2. The voltage value increases as the speed of cutting lines of force increases. An example is the speed of a generator's armature.
3. The voltage value increases with an increase in the number of lines of force.

Example: Increasing the excitation amperes in the shunt field of a DC generator, or the rotor windings of an AC alternator, will increase the voltage value.

The speed at which lines of force cut conductors can be very high when AC or DC is switched off. The arc that forms when the load-carrying contacts open is destructive to the contacts, but the effect of the arc is to slow the shutdown of power. When the shunt field of a DC machine is shut off, the magnetic field from thousands of turns of wire will collapse. As the magnetic field collapses, its lines of force cut the shunt field's turns, producing a very high voltage. If this shutdown is instantaneous (as done with electronic switching), the result is an extremely high voltage spike.

Any sudden voltage change from full voltage to zero or to another voltage will cause the amperes and the resulting magnetic field around conductors to change at the same rate. Any conductor within this magnetic field will have voltage transformed into it. The result can be a very destructive spike that arcs through insulation, and can also destroy electronic circuitry. The voltage spike value increases as the length of the conductors increases. If the distance between an electronic speed controller and the motor it controls is over 50 feet, spike damage to the motor's winding is common. Spike-caused insulation breakdown will occur within the first few turns of a motor's line lead. Spike voltage can be dampened with reactors—a coil of heavily insulated wire wound around laminated iron. A reactor bucks sudden power changes.

Power lines in Canada that feed the eastern United States have problems with magnetic storms from the sun. Although the magnetic field is barely detectable, the voltage becomes very high and destructive when accumulating over hundreds of miles. These storms have moved magnetic North so far that survey crews have had to adjust for it.

Steel conduit that gets hot is another problem caused by magnetism. A conduit containing multiple, unevenly loaded conductors gets hot from eddy current. Eddy current circulates in steel when magnetic lines of force cut this ferrous metal. (This heat is a direct power loss.) If the amperes are the same in each conductor, their magnetic fields cancel each other and reduce the heating effect.

A neutral wire or a bare equipment ground wire can become dangerously energized by being located close to the magnetic fields of other wires. An equipment ground wire (that doesn't go back to the transformer) can be ineffective if connected to a ground rod that doesn't carry enough amperes. In the case of dry sand conditions, it would be safer to use an ungrounded power supply. (Poorly designed electrical systems in desert locations can be deadly where plumbing is involved.)

Troubleshooting Less Common Motors

Troubleshooting Identified Motors

The troubleshooting procedures under "Typical Winding Problems" in Chapter 6 can be used on all types of three-phase motors.

The motor's connection first must be identified. The schematic of the motor then is used to test equal circuits and, if necessary, to locate the problem. Comparison testing is the most accurate and dependable test procedure.

All circuits should be given the proper winding-to-frame test (with the usual precautions), starting with an ohmmeter.

Power supply problems (unbalanced voltage, low or high voltage, spikes, etc.) are all in-plant problems that cause motor failure. These problems don't cause immediate motor failure so they are often overlooked.

Bearing breakdown is high on the motor failure list. Information is found under "Bearing Maintenance," later in this chapter.

Keeping good records on special motors can shorten troubleshooting time. Some motors and controls need more attention than others. Odd characteristics of a motor, such as high amperes of the multispeed consequent-pole motor's low-speed connection, should be noted in the motor's maintenance record. Control contacts that deteriorate sooner than normal should also be noted, as well as solutions to past problems. Having easy access to this type of information can shorten (or prevent) downtime.

Special motors such as multispeed or special-frame motors are expensive and hard to replace. (They usually have to be rewound or repaired.)

Special-duty motors often have copper or other alloys (other than aluminum) in their rotors. Broken rotor bars occur much more often with these alloys than with cast aluminum. Symptoms of broken rotor bars are found in Chapter 3, "Broken Rotor Bars."

Troubleshooting the Synchronous Motor

The brush-type synchronous motor components most likely to break down are

- The DC exciter fields and discharge resistor
- The amortisseur (squirrel cage) winding
- The stator winding
- Bearings

 Caution   The brushes, brush holders, and slip rings can have extremely high voltage while the motor is operating. Use caution with these components.

The DC Exciter Field and Discharge Resistor

The DC exciter field and its discharge resistor are a closed circuit (Fig. 7.1) until the motor control energizes the DC field and then disconnects the discharge resistor.

When the motor starts, the DC field will have high voltage transformed into it. The transformed voltage is controlled by the overloading effect of the discharge resistor. If the resistor develops a faulty connection, it won't load the field circuit enough and voltage will become too high. The DC field may not fail immediately, but its insulation has been stressed and will eventually break down. Most DC field failure is related to high-voltage stress.

A shorted DC field coil may keep the machine from reaching a speed high enough (95 to 98 percent of synchronous speed) to apply the DC voltage. If too many turns are shorted out, a circulating current develops and forms a pole in the coil's iron. This pole will buck the pole formed by the amortisseur (squirrel cage) winding, causing loss of starting torque, and the motor won't be able to get up to speed.

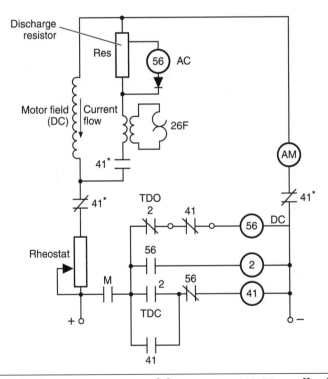

FIGURE 7.1   The function sequence of the contacts 41°. Normally closed contact 41° opens after normally open contacts (also 41°) close.

For the first test, visually check each coil for signs of shorted coils. (The coils are usually wrapped and varnished, making it hard to see signs of a short.)

The DC exciter field can be tested with the same procedures that are used on the shunt field of a DC motor. (The resistance of the field coils controls the current.)

The coils can be comparison-tested using the motor's DC source and a voltmeter (voltage drop test) or by using an ohmmeter. The current should be limited with resistors to about one-half of the field's ampere rating.

For the DC voltage drop test, divide the applied voltage by the number of coils—to get the approximate voltage drop across each coil. The voltage read across the coils should be ± 5 percent of each other.

An ohmmeter or microhmmeter can also be used to compare the resistance of the coils. The allowable difference is ± 2 percent.

The discharge resistor limits the transformed voltage in the DC field each time the motor starts. Its purpose is to overload the circuit and cause a voltage drop, the same as an overload affects the secondary of a transformer.

The resistors are usually made of ni-chrome wire, which expands and contracts each time the motor starts. If it is adjustable, the movable contact (clamped to it) may loosen. The discharge resistor, its control, and all connections in the circuit should be checked at least once a year. Connections can be checked with an infrared gun (as the motor starts).

If a coil is shorted, the rest of the coils in its (series) circuit may be damaged. They should be individually surge-tested for turn-to-turn insulation damage.

The DC field circuit's resistance to ground should be tested yearly with an ohmmeter and a megohmmeter. Early detection of insulation leakage from brush dust or other conducting contaminants will avoid a catastrophic breakdown.

The Amortisseur Winding

The amortisseur (squirrel cage) winding is designed to bring the rotor up to 95 to 98 percent of synchronous speed. At this time DC is applied, bringing the rotor up to synchronous speed.

The amortisseur winding is not designed for continuous duty. Each time the motor starts, the amortisseur winding expands and contracts. This will eventually cause the bars to crack and become open. Open bars cause a loss of torque, and the rotor can't get up enough speed to complete the start cycle. (See "Assorted Rotor Problems" in Chapter 6.)

The amortisseur winding can be checked with an infrared gun (if the motor has an open frame). Start the motor and run it long enough to heat the amortisseur winding. Check the bars and end ring connections immediately. A bar that is cooler than the rest is open. An end ring connection that is hotter than the rest has a bad connection.

Stator Winding

The stator winding is connected in the same way as any three-phase stator. It will be connected either wye or delta—usually multiple wyes or deltas. It's subject to the same problems explained in Chapter 6.

Large synchronous motors are form-wound, with each coil wrapped individually. They are designed to operate on high voltage.

The same test procedures are used on both large and small motors. Instruments for testing them include an ohmmeter, a megohmmeter, a microhmmeter, and a surge tester.

The first test should be done with an ohmmeter, from a lead to the frame of the motor. A low reading (less than 50 megohms) indicates that conducting contaminants may be on the windings. The megohmmeter can be used next to verify the ohmmeter readout. *Be sure to ground a lead to the frame after the megohmmeter test.*

The comparison test for shorted windings can now be done with a microhmmeter at the motor terminals in the control box. (Breakdowns such as shorted coils are usually very visible.)

If there is a difference in the readings of the comparison test (compare leads 1 to 2 with 2 to 3, and 3 to 1), use the surge tester to confirm the microhmmeter results.

Corona

High voltage will cause a phenomenon called corona. Corona is the blue light that surrounds the coils while the motor runs. (The air next to an energized winding becomes ionized and glows.) It will also make a hissing or buzzing sound. Like any electrically caused spark, the corona gives off a radio frequency.

A minute amount of damage to the insulation is done each time the corona discharges. It erodes a tiny particle of insulation which—over time—will produce a fine white dust. (This can be minimized by coating the coils with a conducting paint in the area where they contact the slot iron.) Damage occurs slowly—over years in most cases.

More corona damage is caused by the ozone it produces. An arc changes oxygen to ozone gas. This gas attacks insulation by chemically degrading it. Ozone does the greatest damage in voids in the slot, or between turns. Vacuum/pressure impregnation (VPI) application of varnish will reduce the number of voids. (The varnish must have the right level of viscosity to successfully eliminate voids.)

Bearings

Worn sleeve bearings reduce the air gap on one side of the rotor. The result is extreme mechanical stress to the shaft and bearings. There should be about 35 pounds per square inch of magnetic pull, equally distributed around the rotor (if it is centered). If the air gap is reduced by one-half on one side, the magnetic pull on the wide side will drop to approximately 10 pounds per square inch, while the narrow side will go up to nearly 150 pounds per square inch. Bearing wear is accelerated by the increased stress.

Air gap should be measured yearly on pedestal bearing motors. Temperature change can cause movement of the supporting components. (There have been cases of concrete swelling when reinforcement mesh rusts.)

Step Voltage Test

The step voltage test (explained in the Chapter 8 section, "Instruments for In-depth Testing and Scheduled Maintenance") is a reliable method of testing large synchronous motors and generators. It's a maintenance procedure, done with an instrument called a winding analyzer (Fig. 7.2). The test should be scheduled on a regular basis.

Troubleshooting the Two-Speed One-Winding Motor

Always disconnect and lock out the power when testing a motor that is connected to its control. Test the two-speed motor using the same test methods as with a nine-lead motor. The first test should be from a motor (T) terminal to the conduit or control box with an ohmmeter or megohmmeter. If a ground is indicated, the motor should be disconnected and checked (separately from the conduit). A solid ground means the motor must be replaced or rewound. Fifty megohms or less means the motor should be cleaned and dried.

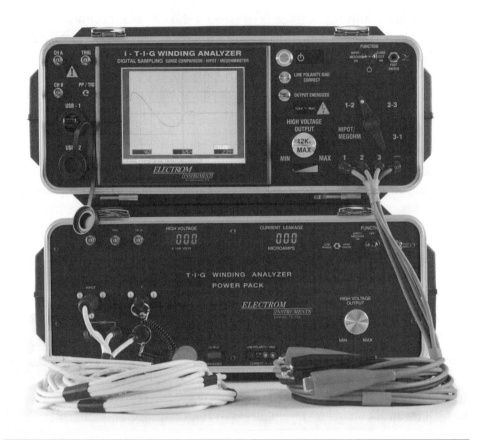

FIGURE 7.2 The winding analyzer is used to perform the step voltage test. *Electrom Instruments.*

If no ground is indicated, the windings can be tested using the line-to-line comparison test. A comparison test can be done from L1 to L2, L2 to L3, and L3 to L1. (It may be necessary to turn the shaft while testing.) The controller contacts can be blocked or held closed to check the two speeds separately. A difference in test results requires further testing.

The numbers in the two-speed winding schematic can be used as a guide for comparison testing. Both constant-horsepower and constant-torque windings have an equal number of turns or coils between T4 and T5, T5 and T6, and T6 and T4. An equal number of turns or coils is also found between T1 and T2, T2 and T3, and T3 and T1.

Any fault that is found should be verified by disconnecting the motor from the controller and testing the motor by itself.

If a surge test is used, run the test on both speeds. This will test all the turns that are connected to the lines. (These turns receive line spikes, and might have turn-to-turn insulation damage.)

The variable-torque motor should be tested in the same way as the constant-torque and constant-horsepower motors. (It may be necessary to turn the shaft during the test.)

Tests should be made from T1 to T2, T2 to T3, and T3 to T1. Another test series should be made with T1, T2, and T3 tied together. The test is now made between T4 and T5, T5 and T6, and T6 and T1.

Troubleshooting Multispeed Multiwinding Motors

Always disconnect and lock out the power when testing a motor that is connected to its control.

The first test is a ground test from each winding to the conduit (Fig. 7.3). The next test should be from one winding to the other. (Be sure they are not connected to a common terminal in the control or internally to a common lead.) Failure of either test requires disconnecting the motor for further testing. Both problems (windings to ground and winding to winding) require rewinding or replacing the motor.

Caution must be used when a test is done that involves voltage. Voltage is transformed into the winding that isn't being tested and could be hazardous.

If the leads of the idle winding touch each other, the test results (on the winding being tested) will be distorted.

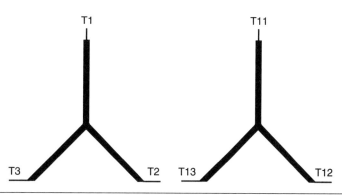

FIGURE 7.3   Schematic for a two-speed two-winding motor.

Each winding of a multispeed multiwinding motor should be tested as a separate motor. The windings normally aren't connected to each other internally. They may, however, use a common terminal in the control. (In this case, the windings should be isolated from each other before testing. The ground [lead-to-conduit] test can be made without the windings being isolated.)

All windings should be tested (with the correct schematic) as described in the preceding section, "Troubleshooting the Two-Speed One-Winding Motor."

Circulating Current in Multiwinding Multispeed Motors

Some motors with three or more speeds require their two-speed winding(s) to be open when not in use. As seen in Fig. 7.4, a circulating current is transformed into the idle winding from the energized winding. The power used by the circulating current increases the amperes of the energized winding and quickly overheats it. To prevent circulating current, the idle winding is opened, as shown in Figs. 7.5a and 7.5b. The lead T7 or T17 is the end of the opened phase. (Lead T17 is connected to the higher-speed winding.) The motor's control joins T7 (or T17) with its respective lead when

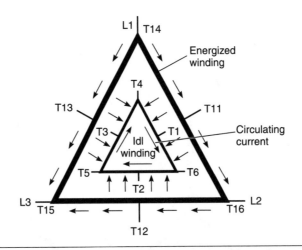

FIGURE 7.4   Schematic of a constant-horsepower winding that is sharing the stator with another winding. A circulating current will occur if the idle winding isn't opened.

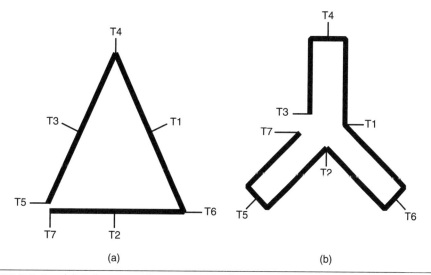

(a)   (b)

FIGURE 7.5   (a) Schematic for a constant-horsepower motor and (b) for a constant-torque motor, connected so they won't have circulating current.

the winding is energized. Lead T7 (or T17) is joined with T3 in the constant-torque motor and with T5 in the constant-horsepower motor.

Some three-speed motors have a two-speed winding and a one-speed winding sharing the slots. The single-speed winding will have a special internal connection that doesn't require the idle two-speed winding to be opened. If this motor isn't connected as it was originally connected, the single-speed winding will get hot (because of circulating current in the two-speed winding).

Circulating current occurs because the poles of one winding align with the poles of another winding. A proper connection will allow another pole (in its circuit) to cancel the circulating current created by an aligned pole.

A winding needs half as many circuits as it has poles, to prevent circulating current without opening it. Each internal circuit must contain two poles that are located on opposite sides of the stator. Spacing two poles of the circuit across from each other keeps them out of magnetic alignment with the poles of another winding (which has a different number of poles). This arrangement cancels current that would circulate because of the closed loop. If both windings are connected single-circuit wye, this spacing isn't necessary.

Rotor Problems in the Multispeed Motor

The multispeed motor has a squirrel cage rotor like any induction motor. It is more likely to have bars made with copper, brass, or a similar alloy. These alloys have more problems than aluminum. Rotor problems include open rotor bars, open end rings, and other problems described under "Function of the Squirrel Cage Rotor" in Chapter 3 (see Fig. 3.48).

Open rotor bars and end rings decrease the motor's power, as described earlier. If a motor takes longer to start, or runs hotter (and more slowly) than it should, the rotor may have a problem. (Low supply voltage will also cause this problem.)

Open rotor bars can be detected with a (correct size) ohmmeter attached to any two winding leads of the same winding. Turn the shaft; note the high and low readings. Departures from normal readings indicate an open rotor bar.

Rotors can be rebuilt to original specifications. They can be rebarred if made of copper or other alloy, or recast if made of aluminum. Multispeed motors are costly and hard to replace, making it cost effective to rebuild their rotors. Rotors of large motors are routinely rebuilt (Fig. 7.6).

FIGURE 7.6 A rotor being completely rebuilt. *Jasper Electric.*

Testing Motors in Place

Test Instruments for Testing in Place

The following test instruments can be used to troubleshoot the nine-lead three-phase motor in place: clamp-on volt/ammeter, ohmmeter, megohmmeter, tachometer, induction-based test instruments, and infrared gun.

The digital clamp-on type of volt/ammeter is capable of capturing instantaneous lockout amperes. The lockout feature records the locked rotor ampere reading.

An ohmmeter and a megohmmeter are used to test slot insulation (winding lead to frame).

The tachometer is used on a loaded motor to see if it's below its nameplate RPM.

Induction-based test instruments, described in Chapter 8, are high-frequency types of testers. They work on motors of any size.

The infrared gun locates hot spots such as poor connections. This instrument has many other uses; some are described in Chapter 8.

Core and Rotor Damage

If a motor has tripped its protection or blown a fuse, the cause of the disconnect should be identified before the motor is reenergized. (A failed motor that has had arcing in its slots would be unnecessarily damaged if the motor is restarted.) If a ground or short occurs in the stator slots, the motor's core and rotor will be severely damaged. Sometimes a large number of slot teeth will fuse together with melted copper. The copper must be ground, filed, or chiseled out of the core before rewinding the motor. In some cases, so many teeth are removed that the core's magnetic balance is thrown off.

If there is extensive core damage, there are three options: If repair is not economical, replace the motor. Restack the laminations, staggering the damaged laminations evenly around the stator. Replace the damaged laminations (Fig. 7.7).

If an arc forms in the slot near the air gap, it's directed at the rotor. This can melt aluminum rotor bars in just a few seconds. Motors up to 100 horsepower with this type of damage may not be (economically) repairable and therefore may require replacement.

FIGURE 7.7 New replacement laminations. *Jasper Electric.*

A motor connected to an ungrounded system is less likely to suffer severe core damage. It may run normally until another part of its winding, another device, or another motor on the system develops a ground.

Troubleshooting the Nine-Lead Motor (Wye or Delta) in Place

If a motor is smoking or has any other obvious major problems, it has to be replaced. The cause of the problem, however, should be identified. If the motor's history is recorded on computer, the data should be checked before troubleshooting the motor.

If someone operates the motor, get that person's input. (This may help identify the problem more quickly.) Factors of concern are load change, modifications in the machine, sound change, and history of past problems. Load change or machine modification may require a larger motor.

Visual Check of the Nine-Lead Motor (Wye or Delta)

If the motor has an open frame, visually check for burned windings. If you can't see the windings, check the paint on the shell of the stator. If the windings have heated excessively, the paint will look discolored or scorched where the core meets the shell. An overheated winding or bearing journal can be easily identified by the discolored paint. (Special paints are available that will change color when they reach a *predetermined* high temperature.)

Check for a burned-winding smell in the motor's connection box if the motor is totally enclosed. Burned windings require rewinding or replacing the motor.

The preceding checks should be made before doing more extensive testing.

Check for open protection (fuse breaker or control protection). If fuses are blown or other protection is open, don't restart the motor. Restarting the motor can damage the service line components and the motor's major components (core or rotor).

Shut off and lock the motor's main disconnect switch so it can't be reenergized. Check the incoming voltage of all three lines with a voltmeter. The supply voltage should be within 10 percent of the motor's voltage rating. (If the motor can run loaded, the voltage should be tested while the motor is running.)

Check for voltage unbalance. An unbalance greater than 1 percent can overheat a motor and cause failure. This problem doesn't cause immediate failure. When the motor does fail, its winding looks like those in Fig. 7.8.

FIGURE 7.8 A winding that failed from unbalanced voltage. *EASA.*

Testing from Control (Wye or Delta)

If none of the visual observations shows a problem, the motor can be tested from its control (or disconnect). The following tests are done from the control.

A standard ohmmeter (multimeter) should be used to test for grounds and opens. It can also be used on the windings of smaller motors (up to 10 horsepower). A microhmmeter is used to test larger motors. Induction-type test instruments will test motors of any size. These instruments are all covered in Chapter 8.

After the power is off and locked, use an ohmmeter to test (at the T [motor] terminals in the control) for the following:

- Ground test from the control
- Open winding test from the control
  - Open winding test (wye)
  - Open winding test (delta)
- Comparison test from the control
- Rotor test from the control

Testing from the control also tests the lines to the motor. A motor that fails any of these tests must be rewound or replaced.

Ground Test from the Control

Disconnect and lock out the power. The ground tests start from the motor side (T terminals) of the control. The first test should be with a multimeter, set on ohms. (Use it first to check for a grounded winding, because it wouldn't have enough power to damage contaminated insulation.)

Test from any motor terminal to the conduit. A short (or pegged) reading means that there's a ground—in the line or in the motor.

Open the motor's connection box and carefully check the motor lead connections for a possible ground to the box or to its cover. Disconnect the motor and test from the leads to the motor's nameplate—or to a clean spot on its frame. If the ground is in the motor, it has to be replaced.

Open Winding Test from the Control

Disconnect and lock out the power. Test from T terminal to T terminal with an ohmmeter. If an open circuit is found, it will be in the line or in the motor winding.

Disconnect the motor leads and check for an open winding.

If the open is found in the line, the motor has run on two lines (single-phased). There's a good chance the motor has winding damage and will look like Figs. 7.9 and 7.10 (see "Burned Windings from Operating on Single Phase" in Chapter 6). Normally, an open winding means the motor must be rewound or replaced.

Open Winding Test (Wye)   Disconnect and lock out the power. An open circuit in a single-circuit wye (high-voltage) connection will test open between the open phase and both normal phases (Fig. 7.9).

If a nine-lead motor is connected low voltage, leads T7, T8, and T9 should be disconnected and tested separately for an open circuit.

Larger motors have more than one internal wye. In this case, the open phase has higher resistance than the other two phases. A microhmmeter or surge tester will identify this problem.

If the motor runs, it has less power, and may not be able to pull its load. The motor's sound may change (depending on its internal pole-to-pole connection) because of the magnetic unbalance within the winding.

Open Winding Test (Delta)   Disconnect and lock out the power. Figure 7.10 shows a complete circuit through two phases. The resistance is higher across T1 and T2 (the open phase) because the circuit includes two phases in series. Both T2 to T3 and T3 to T1 show lower resistance (the resistance of only one phase).

A test light will light between all leads when there's an open circuit in a delta-connected winding, and won't indicate a problem.

If the winding has multiple circuits in each phase (e.g., four and eight delta), the test results will be much the same as in the preceding test.

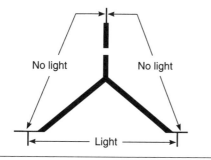

FIGURE 7.9   A single-circuit wye-connected motor with an open phase.

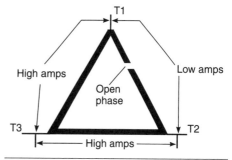

FIGURE 7.10   A single-circuit delta-connected motor with an open phase.

Resistance will be higher across the open phase than across the other two phases. If there's a small difference in resistance, the rotor should be turned to rule out its effect on the test. A microhmmeter or surge tester will identify this problem.

Comparison Test from the Control (Wye or Delta)

Disconnect and lock out the power. Comparison-test with a (correct size) ohmmeter or an inductance-type instrument.

From the motor's control, compare the test instrument readings from T1 to T2, T2 to T3, and T3 to T1. (They should be identical.) If there is just a small difference, turn the motor's shaft slowly at least two complete turns while doing each test.

The rotor bar's position can make a difference in the readings. Note the high and low on all three tests.

A difference in readings between lines indicates a shorted winding or a high-resistance connection between the control and the motor. The motor should be disconnected and tested separately before it's removed.

There's no exact resistance value specified in a motor winding (related to its horsepower and speed) because there are many variations of windings and core length designs.

Two identical test instruments may not give exactly the same readings. This fact should be considered when using a specific test value.

Comparison-testing between winding circuits is the most dependable way to locate winding problems. This can be done quickly on motors of all sizes, and doesn't require manufacturer's specifications.

Rotor Test from the Control

This test should be done if the motor doesn't have normal power (if it takes longer than previously to start the load, or if the shaft speed [loaded] is lower than the nameplate RPM).

Disconnect and lock out the power. The rotor test can be done during the comparison test. When turning the shaft during the ohmmeter or induction test, look for a departure from the high and low readings. If a larger departure happens at the same shaft position, the rotor has an open rotor bar (or bars). If a departure is noted at several equally spaced shaft positions, it could be the rotor's design.

Starting the Motor After the Comparison Test from Control

If the motor isn't open or grounded and the comparison test is even, the motor can be started. Check the amperes of all three lines, and compare the results with the motor's nameplate. (There's always a small difference in ampere readings between lines.) To verify that the motor has no problem, use the procedure described in the following section.

If the amperes are normal, it's important to find out what caused the disconnect to open. If the cause of the problem isn't found, the incident should be recorded for future reference. Possible causes are covered under "When Motors Overheat" in Chapter 6.

Line/Lead Interchange Test for Uneven Amperes

Most three-phase motors will have different ampere readings on all three lines, especially when the motor has no load. If the amperes are quite uneven, the problem lies in the power supply or in the motor winding.

This can be determined by first recording the amperes of each line and then interchanging all three lines or all three motor leads, whichever is easier. (Changing all three keeps the rotation the same.)

If all three lines keep the same readings, the problem lies in the line. If all three motor leads keep the same reading, the problem lies in the motor.

Recording Motor Faults

A history of motor breakdowns is valuable in maintaining production. Future problems are then easily recognized and possibly prevented. Ideally, a plant's motor problems are all recorded on a good software program.

A well-designed report sheet makes recording problems easier. It should include a sketch of a stator showing the location (top, bottom, front, back) of the breakdown. For example, a breakdown that occurs frequently at the bottom of the motor indicates a moisture problem. (Simply drilling a drain hole may be the solution for this particular problem.)

Most failures are visible, but the fault should be recorded, with reference to a lead number.

Schematics of the wye and delta nine-lead connections should be used to reference the location of all breakdowns (according to lead location).

Line conditions should be recorded. These include voltage and amperes of each line (no load and full load). If possible, the readings should be taken both at the control and at the motor.

Locating Motor Faults

Test Equipment for Locating Motor Faults

Locating and identifying motor faults can pinpoint failure causes. This is an important part of preventive maintenance. A motor service center should identify, whenever possible, the cause of failure. Identifying a failure is especially important with motors that are replaced rather than repaired.

Limited single-phase current works very well for testing three-phase stators. Figure 7.11 is a diagram of a test panel that can be constructed with stove elements. The test panel can be designed for higher current. Low-resistance grids can be used if more amperes are needed.

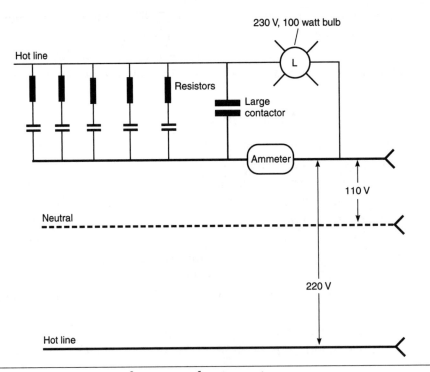

FIGURE 7.11   Circuitry of a test panel, using resistors.

This test panel can also be used with a higher hertz power supply than 60 Hz. (Locate the ammeter on the 60-Hz side of the power supply.) Higher hertz creates more inductance and lowers the amperes and torque of a motor being tested.

High-hertz test instruments create more inductive reactance than do 60-Hz test instruments. A few shorted turns are much more apparent with induction-based test instruments than with an ohmmeter.

An ohmmeter uses low-voltage DC and measures only the resistance of the wire. The short length of wire contained in shorted turns (that are eliminated from the phase) has very little resistance. Identifying a faulty phase by comparing its resistance to two normal phases is harder to do with an ohmmeter.

This type of motor testing should be done in a dry location, because if the winding is grounded to the frame, the test can be dangerous. The motor may have to be disassembled to accurately diagnose and locate the winding fault.

The test results that follow are described as high or low resistance (when using an ohmmeter), or low or high amperes (using limited single-phase current). The term *high resistance* on an ohmmeter is the same as *low current* with limited single-phase current.

Locating Shorted Turns in a Wye-Connected Nine-Lead Motor

In this case scenario, the motor winding being tested is a nine-lead wye connection with shorted turns in phase A. The breakdown isn't visible. It was operating on high voltage when it failed.

First, check the rotor winding by turning the shaft, using an ohmmeter or limited current. Compare the high/low readings as described earlier. (The stator winding can be checked more quickly and easily with the rotor removed, but it isn't required.)

The motor's shorted A-phase winding has less resistance than the B and C phases (Fig. 7.12). Because of the wye connection, there is lower resistance (or higher amperes) from the faulty phase lead (T1) to both of the other phase leads (T2 and T3). When the two normal phases (T2 to T3) are checked, they will have higher resistance (or lower amperes) because the shorted turns aren't included.

If more than one phase has shorted turns, all three phases will have different readings.

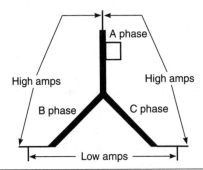

FIGURE 7.12 A single-circuit wye-connected motor with an open phase.

If the short isn't visible, the exact location of the shorted turns can be determined by disconnecting leads T4 and T7. Compare the circuits T1 to T4, T2 to T5, and T3 to T6. If the test results of circuits T2 to T5 and T3 to T6 are identical, but different from that for T1 to T4, the problem lies between T1 and T4.

Now cut the tie cord and carefully lift the T1 lead. If the short is found close to T1 (within the first two to three turns), the breakdown may have been caused by a voltage spike.

If the T1 to T4, T2 to T5, and T3 to T6 test results are identical, compare T7 to T8, T8 to T9, and T9 to T7. Record the lead number that shows a difference to both of the other leads in the circuit. A problem in this circuit wouldn't be spike-related unless the motor had been operating on its low-voltage connection.

Locating Voltage Spike Damage in a Wye-Connected Nine-Lead Motor

Voltage spike damage has increased with the use of variable-hertz drives. This problem can be reduced with properly placed reactors, surge capacitors, or tuned filters. Power quality test instruments can be used to determine the likelihood of spike-causing conditions. Figure 7.13a shows an instrument used for this test.

Spike damage occurrences increase when the distance between the control and the motor is more than 50 feet. This will cause a breakdown within the first two to three turns of a line lead. This breakdown characteristic is unique to spikes and is easy to identify (Fig. 7.13b).

If the motor was operating on low voltage, test each circuit separately, as shown in Fig. 7.14. The leads that connect to lines are T1 and T7 for line 1,

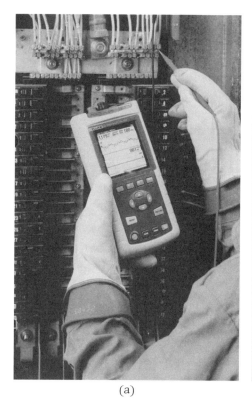

(a)   (b)

FIGURE 7.13   (a) Testing the power quality with a handheld oscilloscope. *Fluke.*
(b) Winding damaged by a spike. *EASA.*

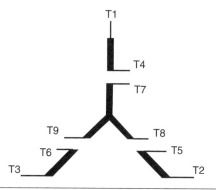

FIGURE 7.14   Check points for finding spike damage in a nine-lead
wye-connected motor (connected low voltage).

T2 and T8 for line 2, and T3 and T9 for line 3. If the damage is not visible, each identical circuit should be comparison-tested (T1 to T4, T2 to T5, T3 to T6, T7 to T8, T8 to T9, and T9 to T7).

A brief description of the breakdown and its location should be entered in the motor's software for future reference.

Locating Shorted Turns in a Delta-Connected Nine-Lead Motor

In this case scenario, the motor winding being tested is a nine-lead delta connection, with shorted turns in phase A. The breakdown isn't visible. It was operating on high voltage when it failed (Fig. 7.15).

First, check the rotor winding by turning the shaft and comparing the high and low readings. (The stator winding can also be comparison-tested with the rotor removed.)

The motor's A-phase winding has shorted turns. This means it has less resistance than the B and C phases (Fig. 7.16). Because of the delta connection, there will be lower resistance (or higher amperes) across the faulty phase leads T1 to T2. When the two normal phases (T2 to T3 and T3 to

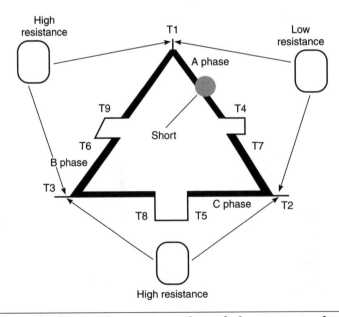

FIGURE 7.15 Checkpoints for locating a shorted phase in a nine-lead delta connection (connected high voltage).

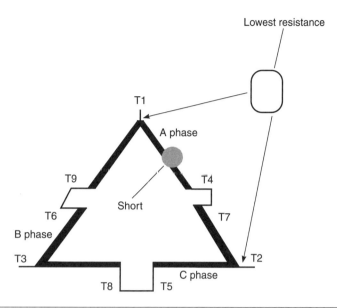

FIGURE 7.16   A shorted A phase in a nine-lead delta connection (connected high voltage) gives the above test results.

T1) are checked, they will have higher resistance (or lower amperes) because the shorted turns aren't included. Some current will flow through the faulty phase when the normal phases are tested. This current will affect both normal phases the same, and the readings will be identical. All three phases will have different readings if more than one phase has shorted turns.

The exact location of the shorted turns can be determined by disconnecting all leads and comparing the identical circuits. Identical circuits are from T1 to T4 and from T1 to T9. They should have the same readings if there are no problems. The same is true from T2 to T5, T2 to T7, T3 to T6, and T3 to T8.

The comparison test shows less resistance (or higher amperes) between T1 and T4, which locates the shorted turns. Check the shorted turns between T1 and T4 for a possible spike-caused breakdown.

If the delta-connected motor is operating on low voltage, a spike-caused breakdown will occur close to any lead connected to a line. The low-voltage connection subjects all leads to spike damage (Fig. 7.17).

The location of the shorted coils may point to the cause of the breakdown. (A current-conducting contaminant that settles on a winding will affect the coils that it covers.)

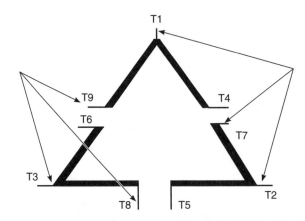

FIGURE 7.17 All leads are subject to spike damage when connected low voltage.

A brief description of the breakdown and its location should be entered into the motor's software for future reference.

Megohmmeter Test for Ground

A megohmmeter (see Chapter 8) is a popular instrument for testing slot insulation. The test is done from a winding lead to the frame. A megohmmeter will show infinity when the winding is good.

If the megohmmeter shows from 2 to 50 megohms to the motor's frame, the motor can still be used, but it should be cleaned and baked dry at the earliest convenience. Follow with a varnish treatment. (The chemical components of new varnish must be compatible with the original varnish.)

Moisture-caused breakdown usually occurs at the bottom side of the lowest coil in the stator. (A water line is often visible on the bottom coils.) This type of breakdown is normally from the winding to the frame (ground) in the lowest slot.

Conducting contaminants cause breakdown in the areas where they accumulate. Breakdowns will occur on top of the top coils, and on the top of the bottom coils.

Be sure to ground the motor leads to the frame after the test.

Bearing Maintenance

Bearing failure causes a high percentage of electric motor breakdowns. Good bearing maintenance is essential for maximum bearing life. The maintenance schedule, the right lubricant, and the amount of lubricant should fit the needs of the motor and its running conditions. (If there is frequent bearing failure, adjusting one or more of these three factors may be all that's necessary.)

Bearings should be replaced if a motor is disassembled for any reason (if it has been in service for a year or more). The cost of new bearings is minimal when compared to the cost of downtime or the cost of replacing or rewinding the motor.

Bearings are manufactured in a sterile, dust- and lint-free, air-conditioned, low-humidity environment. Cleanliness is very important when handling ball bearings. (A grain of sand or a few iron fillings can destroy a bearing prematurely.) It should be wrapped in a grease-compatible plastic as soon as it is removed from the motor's end bracket.

All motor components, tools, and work surfaces should be clean before a new bearing is unwrapped. Immediate motor assembly and bearing installation is recommended. If this is not possible, the bearing should be rewrapped.

A new bearing should never be washed. If the bearing is submerged in oil for heating/installation, the oil must be clean or (preferably) new.

Motor Storage Concerns

Motors should be stored in a dry, clean, vibration-free environment. Humid weather can cause condensation inside a motor. Condensation can be minimized by maintaining the storage room temperature 15 to 20 degrees above the outside temperature.

DC motors develop commutator corrosion in a high-humidity environment. In addition, brush boxes rust, resulting in stuck brushes. Remove the brushes from their boxes and relax the spring tension (for long-term storage).

Turn the motor's shaft a few revolutions at least once a month—daily if there is a vibrating problem. This redistributes the grease and recoats the contact surface between the ball and race.

Ball bearings eventually force their way through grease and make metal-to-metal contact with the races. (If there is vibration, the metal-to-metal contact accelerates, and destructive wear is certain.) The balls wear the races, creating wear spots equal to the ball spacing. This is called false brinelling. Bearings with this problem are noisy and will fail prematurely, even if the motor is new.

Motors in service can also have false brinelling if they sit idle for long periods in a vibrating environment. Turning the shaft or running the motor every day will minimize this problem.

Bearing Failure Causes

Mechanical causes of bearing failure include vibration, excessive belt tension, misalignment, housing or shaft distortion, wrong internal clearance, and preloading from axial thermal growth of the shaft.

Vibration

Many early bearing failures are caused by out-of-balance motor and/or load components. Motors should be balanced to accepted specifications.

A vibration check should be made on the mounting base of a motor. The base shouldn't vibrate more than one-third of the acceptable balance specified for the motor itself. Keeping a history of all vibration tests is recommended.

Excessive Belt Tension

Belts should be tightened only enough to keep them from slipping. Formulas for this recommend about $1/64$ inch of up-and-down belt movement per inch between shaft centers. Worn grooves in the drive pulley require excessive belt tension.

Misalignment

Misalignment is a frequent cause of vibration and preloading-related bearing failure. Although couplings are built to flex and accommodate some misalignment, it should be kept to a minimum. Laser alignment is a very precise method of alignment.

Housing or Shaft Distortion

When the shaft or housing is distorted, the affected bearing race will distort accordingly. The bearing will run hot and will fail prematurely. Check both the shaft and the housing if there is frequent bearing failure.

Wrong Internal Clearance

Bearings have internal clearance that allows for two factors: shaft expansion (from heat) and interference fit (or press fit, which keeps the bearing race from slipping). The extent of shaft (and housing) expansion is related to the motor's enclosure, size, and cooling method. The amount of interference fit varies from motor to motor.

Heat develops in the squirrel cage rotor and travels to the shaft, causing it to expand. The end bracket (end bell) dissipates this heat through the bearing's outer race. (The motor's cooling fan gives the end bracket a different expansion rate than that of the shaft.)

The expansion of the shaft and the pressed fit of the end bracket must be absorbed by the bearing's internal clearance. Motor manufacturers select bearings that fit these requirements. Replacement bearings must have the same specifications as the original ones.

Bearings are manufactured with clearance ratings that allow for expansion (for example, C4). If there isn't enough clearance, the bearing will run hot, causing the grease to fail.

Smaller motors (fractional to 10 horsepower) often have excessive clearance in either the shaft fit or the housing fit. The bearing, which is much harder than either the shaft or the housing, will slip or spin while the motor runs and will wear away material from these components. Over time, a very loose fit develops. This allows the rotor to drag on the stator.

To prevent this problem, an epoxy developed for this purpose can be used (Fig. 7.18). The epoxy remains resilient and allows for expansion, but keeps the bearing race from moving.

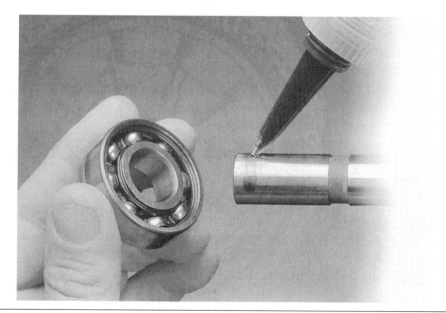

FIGURE 7.18   An epoxy designed to keep a bearing race from rotating. *Locktite Corp.*

Axial Shaft Expansion

Axial expansion occurs as a motor comes up to its running temperature. If the expansion isn't allowed for, preloading becomes a problem. Preloading causes the bearing to overheat and fail prematurely.

The amount of axial expansion is related to the length of the shaft. Thrust washers (made of spring steel) are used to absorb the axial expansion in some motors.

Other motors have hubs that hold the shaft-end bearing captive. The opposite bearing must have room for the shaft's axial expansion. (This can be checked after the motor is completely assembled.) Loosen the hub opposite the shaft. If the hub cover moves outward when the bolts are loosened, the bearing is preloaded. Shims can be used to respace the hub cover. If spring steel thrust washers are used, they shouldn't be completely compressed.

Some motors have a smaller bearing on the end opposite the shaft. Manufacturers do this to cut costs.

Lubrication Schedule and Bearing Life

The lubrication schedule and the type of lubricant are determined by the type of load and the ambient conditions. A belted load needs lubrication more often than a direct-coupled load. Motors that operate under extreme conditions (hot, cold, wet, frequent starts, vibration, etc.) need to be greased more often.

Counter to many grease manufacturers' claims, there is no one grease that fits all conditions. Nor is there a specific lubrication schedule. However, a motor can't be lubricated too often (provided the lubrication is done correctly). Grease scheduling should be frequent enough that the old grease doesn't become caked or hardened. (Forcing hardened grease through the bearing while the motor is running can damage a bearing.)

Motor manufacturers estimate bearing life at up to 100,000 hours if the motor is direct-coupled and around 50,000 hours if it has a belted load. This estimate assumes that a good grease schedule is followed (with the right grease) and that the motor operates in ideal conditions. The speed of the motor, balance (of both motor and load components), and ambient conditions must be considered for individual bearing life expectancy. (Ideal operating conditions are rare.)

Grease Types

Grease is made of various materials that will hold oil. Oil is released from the grease over a period of time. When all the oil has been released, the consistency of the remaining material can range from firm to hard.

Grease types are designed for various conditions. It's important to select a grease type that *fits* the condition. A selection of grease types follows:

- Petroleum oil grease is the most common (with a temperature range of 30° to 300°F).
- Diester oil grease is used for cold conditions as low as –100°F.
- Silicon grease is not recommended for heavy loading but has a wide temperature range (–100° to 300°F). Silicon grease of any type shouldn't be used in a DC motor. Fumes from silicon break down the brush material next to the commutator, resulting in excessive brush dusting.
- Fluorosilicone grease works well where it might become diluted with solvents or chemical contaminants. (It also shouldn't be used in a DC motor.)
- Perfluorinated polyether greases stand up under temperatures as high as 550°F and have good load-carrying capability.

 Mixing different types of grease can cause bearing failure. If they're not compatible, the greases will liquefy or will combine and become thick. In either case, the bearing lubrication won't be effective. Figure 7.19 shows compatible and noncompatible greases.

Changing Grease Types

The right way to change incompatible grease types is to remove the bearing and wash it thoroughly with solvent, before filling it with new grease. However, in most cases, this procedure takes too much time.

The method that follows, under "Lubrication Procedure," is recommended, with the following modifications. Flush the old grease for a longer time than described (to remove as much old grease as possible). Grease the motor using this procedure at least three times (at no more than one-week intervals).

Grease compatibility chart I = Incompatible C = Compatible B = Borderline	Aluminum complex	Barium	Calcium	Calcium 12-hydroxy	Calcium complex	Clay	Lithium	Lithium 12-hydroxy	Lithium complex	Polyurea
Aluminum complex	X	I	I	C	I	I	I	I	C	I
Barium	I	X	I	C	I	I	I	I	I	I
Calcium	I	I	X	C	I	C	C	B	C	I
Calcium 12-hydroxy	C	C	C	X	B	C	C	C	C	I
Calcium complex	I	I	I	B	X	I	I	I	C	C
Clay	I	I	C	C	I	X	I	I	I	I
Lithium	I	I	C	C	I	I	X	C	C	I
Lithium 12-hydroxy	I	I	B	C	I	I	C	X	C	I
Lithium complex	C	I	C	C	C	I	C	C	X	I
Polyurea	I	I	I	I	C	I	I	I	I	X

FIGURE 7.19 Compatibility grease chart.

Lubrication Procedure

High-pressure greasing equipment should not be used on motors. Nor should grease be forced into the bearing at a fast rate with a hand-operated grease gun. That would force grease through the bearing seal and onto the winding of the motor.

Grease won't break down the motor's insulation. However, it does impede the dissipation of heat from the winding. If grease gets into the air gap between the rotor and stator, the motor becomes excessively loaded.

If the motor has grease fittings, called zirks, a purge plug should be at the bottom of the bearing enclosure. Remove the purge plug and clean the grease zirk thoroughly. Run the motor, and pump the grease through the bearing slowly, until about half an ounce of new grease comes out of the purge hole. The motor should run at least 2 hours before the purge plug is replaced. (This allows the bearing to expel excess grease.)

A bearing should be filled when hand-packing it with grease. But the motor's bearing housing should be filled to only about one-third. This gives the bearing room to expel excess grease. If it can't expel excess grease, the grease will churn, causing it to release its oil too soon. (All grease types are designed to release oil over a long period of time.)

Shielded (or sealed) bearings should be used if there is no purge plug. Replace the grease zirk with a plug, and tag the motor to prevent installation of another grease zirk.

Alternate Lubrication Methods

If the motor is in a remote location or needs lubricant at frequent intervals, self-actuated grease cups can be installed. They're available in battery-operated and gas-operated types and are designed to apply lubricant daily, weekly, bimonthly, or monthly. Extreme temperatures may affect the timing of the gas-operated types.

Oil mist is a method of lubrication that works well on high-speed machines. The oil is atomized, and is delivered with a mixture of air to the bearing under pressure. With the bearing enclosure pressurized, moisture and other contaminants are kept out. Normally the oil mist is applied for a short time before starting (to make sure the bearings aren't dry).

Oil injection is another effective way of lubricating bearings. A measured amount of oil under pressure is squirted directly into the bearing.

Gearbox Lubrication

Gearboxes have an oil level that should be checked at regular intervals. Oil change should be done once a year under normal conditions. If the operating temperature goes above 200°F, oil change every 3 months is recommended. An infrared gun can determine the running temperature of the gearbox.

High humidity and wide temperature swings cause condensation inside the gearbox. The oil will become a whitish color when extreme condensation is a problem.

Particle Analysis

A regular program of particle analysis is recommended for gearboxes, which are vital to the operation of an industry. Wear particles from each component can be identified as well as any outside contaminants. Once a record of wear has been established, any abnormal increase is easily detected. Repair can then be done before a catastrophic failure occurs.

Particle analysis is also done with grease. This service can be used to solve frequent bearing failure problems.

Particle analysis is described in detail in Chapter 8.

Shielded Bearings

Single-row bearings with shield(s) or seal(s) are used in most standard-duty electric motors. (Shields are used more often than seals.) A shield retains grease well, but does not keep out all contaminants. Inactive grease (next to the shield) keeps out some (but not all) types of contaminants.

Some motors have bearings with only one shield (facing the rotor) for retaining the grease. If grease is forced into this bearing too fast, it will go through the shield and onto the motor's winding.

Sealed Bearings

Sealed bearings have flexible seals that rub the inner race. The seal has a small amount of drag that decreases the motor's efficiency slightly. If a motor has to operate where sealed bearings are needed, efficiency can't be a factor.

Seals keep out fine dust and contaminants of that nature. They are available in single, double, and triple seals.

If it isn't possible to keep a timely lubrication schedule, the use of shielded or sealed bearings is recommended.

Labyrinth Seal

The labyrinth seal is a noncontact seal that keeps contaminants out of bearings very effectively (Fig. 7.20). High-speed applications need a noncontact type of seal.

Contaminants do several direction changes after entering the seal. When they get to a cavity between the rotor and stator, they are expelled through a purge hole. Any contaminant that isn't expelled is held by centrifugal force until the shaft stops. At this point, an O-ring, held by centrifugal force, pulls together and seals off the path where the contaminant could have entered the bearing. (The movable O-ring also seals in any bearing lubricant that would flow out.) At rest, the motor is hermetically sealed.

FIGURE 7.20   A noncontact seal that keeps most contaminants out while keeping the grease in. *Inpro/Seal Co.*

Labyrinth Seal Construction

The labyrinth seal has two major parts: a rotor (Fig. 7.21a) and a stator (Fig. 7.21b). Both parts have O-rings that isolate the motor bearing from outside contaminants.

The motor housing must be machined to accept the stator. The stator has an O-ring and a press fit. The O-ring and the tight pressed fit combine to seal the stator.

The rotor has two O-rings as well as a pressed fit. One O-ring seals the shaft from outside contaminants and is stationary. The other O-ring is movable and contacts both the stator and the rotor when the motor is at rest, hermetically sealing it. When the motor runs, this O-ring is pulled away from the stator by centrifugal force. It is held away from the stator by centrifugal force in a space machined into the rotor. There is no contact between the stator and the rotor shaft while the motor runs. Consequently, there's no wear or efficiency loss.

(a)

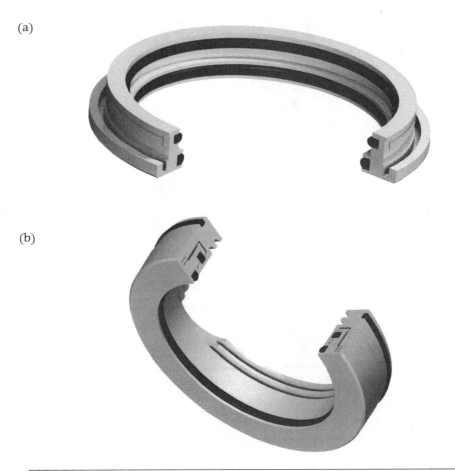

(b)

FIGURE 7.21 (a) The labyrinth seal (rotor). *Inpro/Seal Co.* (b) The labyrinth seal (stator). *Inpro/Seal Co.*

Many motor manufacturers use the bearing isolator (labyrinth) seals in their premium motors. This seal allows them to extend the bearing life warranty. The motor's efficiency is not affected.

Bearing Removal

Disassembling a motor (for bearing removal) must be done carefully. First, remove any bolts or screws that hold the bearing captive. Then remove the bolts (or through-bolts) that secure the end brackets. Loosen the end bracket with a soft hammer (plastic shot-filled or a lead hammer) and a dull chisel. Some end brackets have ears on both sides for this purpose. If there's a deep flange fit, tap both sides of the end bracket uniformly. *Avoid using a steel hammer.*

Bearings should be removed by pulling from the inner race. Very few motors, however, leave access to the bearing's inner race, so a bearing puller has to be used on its outer race. (This is why bearings should be replaced if they are removed.)

A steel washer between the point of the puller and the shaft will keep the shaft's centering hole from being damaged.

If a bearing has spun on the shaft—and seized—when it failed, it may be necessary to use a torch to remove the inner race. This should be done without damaging the shaft. The shaft should be checked with a micrometer and, if necessary, repaired (to its original dimensions).

If there is no shaft damage, the inner race can be removed with an aluminum heating ring (Fig. 7.22). Remove the outer race and bearings, and wipe the inner ring clean. The heating ring is heated to around 500°F, then clamped around the inner race. The race will loosen enough for removal.

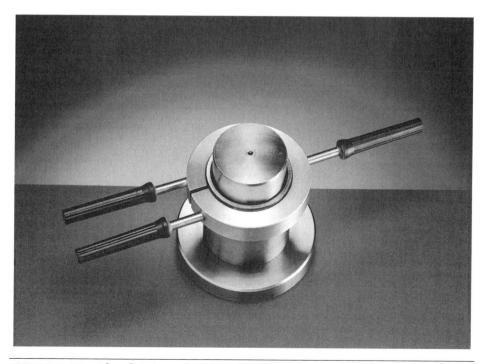

FIGURE 7.22   The aluminum heating ring for removing the inner race of a ball bearing. *SKF USA Inc.*

Bearing Installation

One of the most important requirements for bearing installation is cleanliness. This includes tools, the motor components, and anything that has contact with the bearing.

Small bearings can be installed with a tube that fits the bearing's inner race, and a soft hammer.

A hydraulic bearing press can be used for larger bearings. Pressure has to be *uniformly applied* to the bearing's inner race only.

Heating is a popular method of bearing installation. *Never use an open flame.* The bearing should be heated uniformly and slowly. If the inner race is heated too fast, its rapid expansion will damage the bearing's components, resulting in a noisy bearing.

Overheating is a major concern when any heat method is used to install bearings. The bearing should not be heated over 230°F.

A properly heated bearing should be pushed in place on the shaft without stopping. The inner race will shrink and stick to this spot. If it isn't in place, it will have to be pressed the rest of the way. When a bearing is installed using heat, it will shrink axially a minute amount. Rarely would this be a concern in an electric motor.

Hot oil is a good way to heat a bearing, if done properly. It uniformly heats the bearing, and protects it from rust for a short time.

The oil must be clean and should have a flash point above 480°F, to reduce fire hazard. Also, care must be taken not to overheat the oil because of possible acid buildup.

The hot oil method works best with open bearings that are greased after installation. If the bearing is shielded, oil may dilute the grease.

A heating cabinet is a good method of heating a bearing. The bearing can be kept at the proper temperature in a clean environment for as long as needed using this method.

Figure 7.23 shows an automatic induction bearing heater. This unit goes through a number of important steps. A magnetized temperature sensor is placed on the bearing's inner race. Power is automatically applied intermittently, allowing the heat to migrate to the outer race, which uniformly heats the entire bearing. Once the right temperature is reached, the bearing is demagnetized. Some types of induction heaters don't magnetize the bearing.

The problem with most heating methods is the time and attention involved.

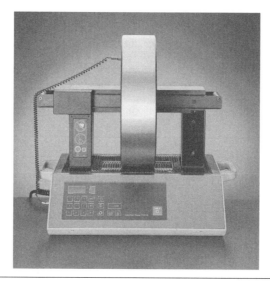

FIGURE 7.23 An automatic induction bearing heater. *SKF USA Inc.*

Identifying Bearing Breakdown Causes

If a motor has frequent bearing failure and the cause isn't obvious, an effort should be made to discover the cause.

The best time to identify the cause of failure is when the bearing begins to get noisy. Mark the bearing's outer race, the housing, the inner race, and the shaft (for reference) while disassembling the motor. This provides a reference point—that is, direction of belt load pull. The bearing can now be broken open to determine the cause of its failure. Some bearing companies provide pictures of failed bearings and identify reasons for their failure.

Bearing Current

Signs of current-caused bearing failure include uniform fluting pits across the inner and/or outer race and inline pits evenly spaced on the balls (Fig. 7.24).

When a voltage potential exists between the shaft and the stator, the current path is through the bearings. The voltage is usually low, and the lubrication between the bearing components has resistance that stops current flow.

There are times, however, when there is metal-to-metal contact. Current flows at this time between the races and the balls, forming a circuit between the shaft and stator.

FIGURE 7.24 The marking found in an electric current–caused bearing failure. *SKF USA Inc.*

There is no damage to the bearing when current flows through it. (Damage occurs when the circuit is broken.) An arc forms between components at this time. (The same kind of arcing occurs when switch contacts open, and break a current-carrying circuit.) Some of the metal vaporizes, leaving pits. Eventually axial lines (fluting) form across the races. Pits form in the balls, the bearing becomes noisy, and it soon fails.

Bearing current breakdown was a rare problem in the past. Variable-frequency drives are linked to an increase in this problem.

Attempts at solving this problem include using current-conducting grease; insulating the bearing from the housing; grounding brushes from shaft to stator; coating the outer race of the bearing with a hard, high-resistance material; and using ceramic balls.

Conductive grease is one of the least successful. The grease has to maintain a conductive path large enough for all the current to flow between the shaft and the end bracket. The active grease is very thin—between the balls and races. Supplying enough grease to ensure good conduction would cause the grease to churn and overheat.

Insulating between the bearing and housing can lower the mechanical strength of a motor.

A grounding brush has been used, with varying results. It must be made of a very low-resistance material (with a high metal content for conducting the low-voltage-driven current). The wear factor between a high-metal-content brush and shaft will be high. A small amount of resistance between the brush and shaft sends current through the bearing, which is a parallel circuit.

A bearing with its outer race coated with hard insulation works well. Such bearings are interchangeable with standard-size bearings and require no alteration of the motor housing. Bearings with ceramic balls also work well.

In extreme cases of bearing current, a combination of a grounding brush and coated bearings can be used. The grounding brush will conduct higher-voltage-driven current that can bypass the bearing's coating. The bearing's coating will stop lower-voltage-driven current that the brush can't conduct.

The best solution is to eliminate the destructive voltage.

Bearing current problems occur in DC motors when their stator field is grounded, or when the power supply is grounded. Bearing current is also a problem with DC motors that are mounted on an electrically common machine, such as a locomotive.

Specialized Bearing Tools

There are many specialized tools for bearing handling. Proper use of them will prevent damage to the bearings, shaft, and outer housing. There are many cases of early bearing failure linked to faulty installation procedures. Training courses offered by bearing companies are highly recommended. The courses teach correct procedures with specialized tools and timesaving methods.

Bearing pullers are available in many variations. Mechanical pullers for small to medium-sized bearings have two or three jaws for gripping the outer ring of the bearing (Fig. 7.25).

Hydraulic-powered tools are used for larger bearing removal and installation. A hydraulic press is also used for removal and installation.

Oil injection is used for bearing removal. The shaft is modified as shown in Fig. 7.26. Oil is forced between the bearing's inner race and the shaft as the bearing is pulled off.

Hydraulic nuts are a tool used to mount bearings. Roller bearings can be mounted precisely on a tapered shaft, using this tool. This type of bearing *must be* mounted precisely for maximum life.

FIGURE 7.25 One of many types of bearing pullers. *Industrial Pulley Puller.*

FIGURE 7.26 The oil injection method of bearing removal. *SKF USA Inc.*

Sleeve Bearings

Most fractional-horsepower single-phase motors have sleeve bearings. The bearings are lubricated with oil-impregnated wicking, which lasts for the life of the motor. Most of these motors have no allowance for lubrication, but can be taken apart to apply oil directly into the oil reservoir. Holes for oiling can be drilled into the bearing cap or the end frame. These motors are usually replaced instead of repaired.

Larger motors with sleeve bearings have an oil ring and an oil reservoir. (The volume of oil needed for large bearings can't be supplied with a wick.) The oil ring rides directly on the shaft, with about one-third of the ring submerged in oil. As the shaft turns, the ring turns with it, carrying oil to the bearing. Oil lubricates all parts of the bearing by flowing through slots along the length of it, and back to the reservoir.

Sleeve bearings in large high-speed motors should be well oiled before they are test-run with no load. This prevents them from being destroyed (wiped) before the oil ring supplies oil.

It's normal to have about $\frac{1}{2}$ inch of endplay in larger sleeve bearing motors. When the motor runs, the rotor will seek its magnetic center. (The shaft shoulder shouldn't ride against the bearing.)

The oil level should be checked once a month, and the oil changed once a year. If contaminants or high temperatures exist, the lubrication schedule should be altered accordingly.

Unlike single-phase motors, three-phase motors with worn sleeve bearing don't show signs of the problem (by becoming noisy). To check for wear, move the shaft back and forth in the direction of the load. There should be no lateral shaft movement.

Chapter 7 Review

1. A failing connection in the DC field/resistor of a synchronous motor's control circuitry will cause the DC field to break down (page 324). T___ F___

2. A shorted DC field coil can prevent the rotor from reaching pullout speed (page 324). T___ F___

3. The amortisseur (squirrel cage) winding (if accessible) can be checked with an infrared gun (page 325). T___ F___

4. A warm end ring and cool rotor bar is the sign of a broken rotor bar (page 326). T___ F___

5. The ohmmeter should always be the first instrument used to test from winding to frame (page 326). T___ F___

6. The vacuum/pressure impregnation (vpi) varnish method is the preferred method of coating and filling all voids in a form-wound winding (pages 326–327). T___ F___

7. An uneven air gap can accelerate bearing wear (page 327). T___ F___

8. The line-to-line comparison test is done first when testing a two-speed motor (pages 327–328). T___ F___

9. The surge test should always be applied to both speeds when testing a two-speed motor (pages 329–330). T___ F___

10. A multiwinding motor is tested as though it is two different motors (pages 329–330). T___ F___

11. When testing a multiwinding motor, the winding that is not being tested has no effect on the test being done to the other winding (pages 329–330). T___ F___

12. The lead T7 (or T17) is not needed in some two-winding, three-speed, three-phase motors (pages 330–331). T___ F___

13. The lead T17 is used on the (pages 330–331)
    a. low-speed winding.
    b. high-speed winding

14. Defective rotors can be completely rebuilt (page 332). T___ F___

15. In some cases, a failed motor can be severely damaged if it is reenergized (page 333). T___ F___

16. Extensive core damage can be repaired or replaced (page 333). T___ F___

17. Operator input is an important part of troubleshooting (page 334). T___ F___

18. Visually checking for obvious signs of catastrophic problems should be done before using test instruments, and often saves time (pages 334–335). T___ F___

19. Checking the voltage and amperes while the motor is operating—fully loaded—should always be done after a motor is installed (page 335). T___ F___

20. Many serious motor faults can be detected at the control or disconnect (page 336). T___ F___

21. Testing all circuits from winding to frame should be done (first) with an ohmmeter. If the motor fails this test, it may need to be replaced or rewound (page 336). T___ F___

22. Open-winding symptoms are the same whether the motor is connected wye or delta (pages 336–337). T___ F___

23. A comparison test—at the motor control—should be done before energizing a newly replaced motor (page 338). T___ F___

24. What are two signs of rotor problems (page 338)?

25. When is the line/lead interchange test done (page 339)?

26. A preventive maintenance program must include the reason a motor failed (pages 339–340). T___ F___

27. The chance of spike damage from variable-hertz drives will increase when the motor is over 50 feet from the drive (pages 342–343). T___ F___

28. If there is a short in one phase of a delta-connected motor, that phase will have less resistance than the other two (pages 344–345). T___ F___

29. The megohmmeter can be used to comparison-test a three-phase winding (page 346). T___ F___

30. Bearing failure causes a high percentage of motor failures (pages 346–347). T___ F___

31. It's important to clean a new bearing before installing it (page 347). T___ F___

32. A new motor that has been stored where there is vibration can have noisy bearings (false brinelling) (page 347). T___ F___

33. If a stored or seldom used motor must be in a vibrating environment, false brinelling can be prevented by spinning the shaft on a regular basis (pages 347–348). T___ F___

34. Belt tension should allow movement of $^1/_{64}$ inch per foot between shaft centers (page 348). T___ F___

35. A ball bearing's internal clearance should allow for interference fit and heat expansion (pages 348–349). T___ F___

36. If a ball bearing's interference fit is too loose—on the shaft or the housing—the component must be altered to eliminate slipping (page 349). T___ F___

37. The amount of axial expansion is related to the outside diameter of the bearing (page 350). T___ F___

38. Grease is available for many types of adverse environments (pages 350–351). T___ F___

39. A good lubrication schedule is once a year (page 350). T___ F___

40. Mixing different types of grease always causes bearing failure (page 351). T___ F___

41. Bearings must be replaced when changing to an incompatible grease (page 351). T___ F___

42. High pressure should be used to make sure enough grease gets to the bearing (page 352). T___ F___

43. Name three alternate lubrication methods (page 353).

44. Gearbox oil should be changed according to the operating conditions (page 353). T___ F___

45. Preventive maintenance for gearboxes should include particle analysis (pages 353–354). T___ F___

46. What is the difference between shielded and sealed bearings (page 354)?

47. The labyrinth seal uses centrifugal force to expel contaminants (page 355). T___ F___

48. If a ball bearing is removed from a shaft, it should always be replaced (pages 356–357). T___ F___

49. A bearing should not be heated over 230°F (page 358). T___ F___

50. The time to identify the cause of frequent bearing failure is when it starts to become noisy (page 359). T___ F___

51. Bearing current failure is caused by current flow interruption within the bearing (pages 359–360). T___ F___

52. Worn sleeve bearings aren't always noisy in three-phase motors (page 363). T___ F___

Chapter 8

Test Instruments and Services

Test instruments are vital to electric motor maintenance and troubleshooting. Lost production makes reactive maintenance ("wait until it smokes") much too costly. Proper application of the right instrument, combined with a good understanding of electric motors, results in quick, accurate decisions.

A reputable organization such as the Electrical Apparatus Service Association (EASA), a worldwide organization of electric motor repair centers, should be consulted before investing in expensive test equipment. An instrument salesperson must be able to show the Institute of Electrical and Electronics Engineers (IEEE) or National Electrical Manufacturers Association (NEMA) standards that verify the values, specifications, and pass/fail limits that are failure indicators. Many of the so-called limits don't exist.

Services that require special training, such as balancing, surge testing, and particle analysis, are offered by electric motor service centers and bearing companies. Many industries outsource these services rather than train their own personnel.

The first part of this chapter covers the first instruments used to troubleshoot a motor problem. The second part is on instruments and tests that are sometimes outsourced.

Instruments for Initial Testing

Instruments for initial testing include

- Voltmeter (with infrared temperature capability)
- Ammeter (clamp-on)

- Ohmmeter (multimeter with built-in infrared thermometer)
- Microhmmeter
- Tachometer (with vibration meter)
- Megohmmeter (megger)
- Infrared gun
- Stethoscope

Voltmeter

The voltmeter (Fig. 8.1) is usually the first troubleshooting instrument used. Even if the problem is obvious, such as a smoking motor, checking the voltage may reveal the reason for a motor's failure. In addition, it verifies that the power is off (for safety concerns).

FIGURE 8.1   The voltmeter/infrared thermometer is the first instrument used when troubleshooting. *EXTECH Instruments.*

A voltmeter tests for line-to-line comparison at the motor's disconnect switch and at the motor's leads (preferably under load). It tests for voltage unbalance (unbalance should not be over 1 percent), an open line, and high or low voltage.

Voltage unbalance will cause a motor to run hot—even if it is not fully loaded—and eventually fail. It's caused by faulty connections, faulty incoming power, and severe upstream load unbalance. A blown fuse will show a voltage reading across it.

This voltmeter has infrared temperature capability. Infrared should be used in the initial test. Check the pulley for heat caused by excessive slippage/worn grooves, shaft end bearing journal, and the stator. This information is valuable in determining the breakdown cause.

Three-phase motors won't start when there's an open line. Their power drops to about half if they are running when the line opens. The open line may show a voltage reading because voltage is fed back through loads that are online. It can be verified by opening the suspected line. Remove the fuse or open the disconnect. There will be no change in the (load side) voltage reading if this is the open line.

High voltage (above 10 percent of the motor's rating) can be caused by power factor correcting capacitors that are online when they aren't needed.

Power suppliers sometimes raise the voltage to compensate for voltage drop caused by a maximum load. High voltage develops when the load is switched off.

Low voltage causes include inadequate wire size (too much voltage drop after load is applied), undersized transformer, wrong voltage (more than 10 percent below the motor's rating), severe load unbalance, and overloaded main feeder circuit.

AC and DC voltmeters for troubleshooting DC machines are covered in Chapter 2. The voltmeter compares the voltage drop across coils (shunt field, series field, and interpoles), reads from brush to commutator (contact drop), checks interpole polarity, checks for continuity between shunt and series field coils, and measures bar-to-bar voltage.

Clamp-on Ammeter

An ammeter (Fig. 8.2) is used on a motor that runs hot and is noisy (excluding bearing noise), and on a motor that starts more slowly than normal. It also tests for line-to-line current unbalance.

FIGURE 8.2 The clamp-on ammeter (with ohmmeter function) gives amperes without touching any live conductors. *EXTECH Instruments.*

Causes of high ampere reading include overload, faulty bearings, shorted winding, open winding, faulty rotor, and voltage problems (covered under the preceding section, "Voltmeter").

A digital ammeter set on high-ampere lockout can be used to estimate locked rotor current (see "Code Letter" in Chapter 5).

The ammeter should always be used on a replacement motor following installation. The best place to take an ampere reading is at the motor leads, under full load.

Ohmmeter

 An ohmmeter should never be used when there is a voltage potential in the circuit. *Line voltage can destroy most ohmmeters.*

The first test (after determining that the power is off) is to test the motor at the disconnect for a ground. If there is a ground, no further testing is necessary from that point. The next step is to disconnect the motor to verify that the ground is in the motor.

The ohmmeter has its own power source (batteries) and uses very low current. Insulation-piercing needle probes can be used because of the instrument's low current (Fig. 8.3). However, an ohmmeter can destroy some electronic components if applied incorrectly.

FIGURE 8.3   A multimeter with an ohmmeter function. *EXTECH Instruments.*

An ohmmeter is used to test a motor's insulation between its winding and its frame. It will detect conducting contaminants without damaging the insulation. It should always be used before applying a high-potential tester or any type of high-voltage testing equipment.

Other uses within its size range include comparison-testing of individual coils or identical circuits; checking an armature for shorted, open, or grounded circuits; and testing a squirrel cage rotor for open bars. Procedures for these tests are described in other parts of this book.

When a good capacitor is tested, the ohmmeter shows a low-resistance reading, which rises slowly until the capacitor is charged. This test indicates the capacitor has capacitance. However, it doesn't give the amount of capacitance. *No* reading means that the capacitor is open. A steady low-resistance reading indicates that it's shorted.

An ohmmeter's test limit for three-phase induction motors is about 10 horsepower. Larger motors can be tested with a microhmmeter.

Microhmmeter

The microhmmeter can test very low resistance. The instrument uses up to 5 amperes (supplied by batteries or a single-phase source). The microhmmeter in Fig. 8.4 measures down to 0.01 milliohm.

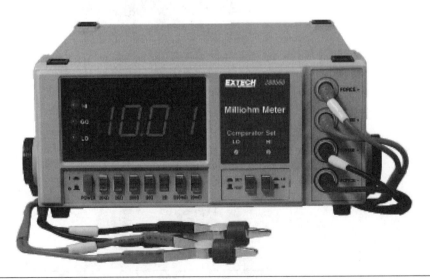

FIGURE 8.4 A microhmmeter designed for low-resistance testing. *EXTECH Instruments.*

Temperature index is an important factor in calculating low resistance and must be used for exact measurements.

When low resistance is being tested, the test probes must have a secure connection to the circuit conductors.

The microhmmeter can be used in the following tests:

- Polarization index
- Testing of large three-phase motor windings, internal connections, and rotors
- Transformer windings
- Bus bar connections and circuit breakers
- Lead connections (contaminant between the lugs)
- Low-resistance coils (interpoles, series field coils, compensating windings)
- Large armature circuitry (riser weld integrity, equalizer connections, and coil resistance comparison)

Tachometer

Handheld tachometers need a piece of tape on the shaft for a reference point. The shaft key will work, although it may have to be filed or scraped clean. Two reference spots will double the reading. Like any test instrument, it takes some practice to develop confidence in it.

A hot-running motor should be tested with a noncontact type of tachometer (Fig. 8.5) before using electrical test instruments. All induction motors get hot when they run more slowly than their nameplate speed. (Loaded RPM shouldn't be lower than the nameplate RPM.)

When a motor's nameplate RPM is close to synchronous speed (less than 1 percent), reduced RPM becomes very critical. Low-slip induction motors (design A and most high-efficiency motors) have a substantial increase in amperes when loaded only a few RPM under their nameplate rating. *Excessive heat* shortens insulation life and bearing life.

Several factors cause a motor to run more slowly than the nameplate RPM. These include overload, rotor problems, operating on low voltage, unbalanced voltage, a shorted or open winding, and an open supply line (single-phased).

FIGURE 8.5 A noncontact type of tachometer (used with nameplate RPM) can quickly tell if a motor is overloaded. This model has an infrared thermometer feature. *EXTECH Instruments.*

Megohmmeter

The megohmmeter (megger) is used to test the insulation between the winding and the core of a motor (Fig. 8.6). It can also be used to test the insulation between windings, such as with the series and shunt field in a DC motor. It can be used for a phase-to-phase test on three-phase motors that have wye-delta and 12-lead connections. (Phases can be isolated with these connections.)

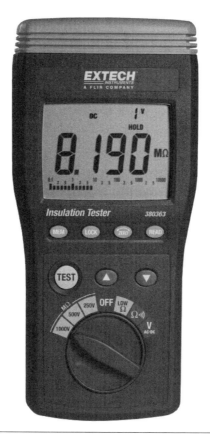

FIGURE 8.6   The megohmmeter. *EXTECH Instruments.*

The megohmmeter produces a DC voltage (500 to 5000 volts) with very low amperes. The test involves time—up to 3 minutes. During this time, the capacitance of the machine is charged.

When two conductors are separated by insulation, they form a crude capacitor (see Chapter 4). The two conductors are the plates of the capacitor. The conductors (plates) are separated by the slot insulation, which is the dielectric.

The winding in a motor has a large area when viewed as a capacitor plate. The slot area of an AC motor (and the pole iron of a DC motor) forms a sizable plate area.

The amount of time it takes to charge the capacitance to full value depends on the amount of capacitance the machine has, and the power output of the megger. There will be a small amount of leakage (measured in megohms) when the full charge is reached. This value should not be less than 50 megohms. (One megohm was originally acceptable.) A new or rewound motor will test infinity.

At the start of the megohmmeter test there's an inrush of current that charges the capacitance of the motor. When the test is completed, a charge may remain between the winding and the motor's frame. For personal safety, the leads should be grounded to the motor's frame to discharge the voltage (after testing).

The megohmmeter is also used in the polarization index test. The procedure is found in the upcoming section "Instruments for In-depth Testing and Scheduled Maintenance."

Infrared Gun

An infrared gun (Fig. 8.7) can detect the temperature of a device without having physical contact. This is useful for checking (or comparing) the temperature of connections, especially high-voltage connections.

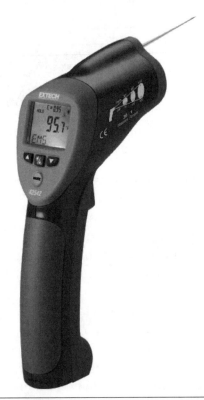

FIGURE 8.7   The infrared gun is a noncontact instrument for detecting temperature-related problems. *EXTECH Instruments.*

A very common, destructive problem for motors is the high-resistance (hot) connection. Every time a motor starts, high inrush amperes (sometimes over 500 percent of nameplate amps) will stress all the motor circuit's connections. Connections should be checked with an infrared thermometer routinely, especially if the motor starts frequently. A hot connection should be checked for oxidation, and cleaned or replaced if necessary.

High temperature usually indicates a motor problem. (Windings or bearings often get hot just before they fail.) The temperatures of like components in an open motor, such as individual brushes or coils, can be compared quickly and safely.

Other uses for this instrument are to check liquid level in tanks; check above-normal heat in pulleys and belts, couplings, gearboxes, pillow block bearings, and (hot) conduit; and check anything hot enough to cause injury.

Some infrared guns have information storage capability and are compatible with computer software programs. The data can be easily referenced when abnormal operation is suspected.

Stethoscope

A stethoscope is used to pinpoint noise sources. Early detection of bearing failure can be done with this instrument. The electronic model (Fig. 8.8) will record information for comparison and trending. It also allows personal observation. The stethoscope also checks for leak detection, liquid flow or no flow, and arcing.

Instruments for In-depth Testing and Scheduled Maintenance

Instruments for in-depth testing and scheduled maintenance include

- Surge tester
- Rotor test
- Polarization index test
- Step voltage test
- High-potential test (hi-pot)
- Power quality test instruments
- Power factor meter
- Vibration monitoring sensors

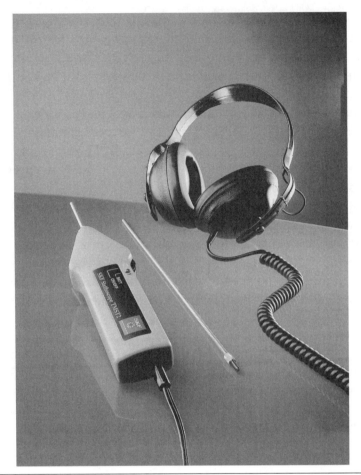

FIGURE 8.8 A stethoscope used to pinpoint problems. *SKF USA Inc.*

- Laser alignment
- Particle analysis

Surge Tester

The surge tester (Fig. 8.9) has the capability of detecting a turn-to-turn fault in individual coils before it becomes an actual breakdown. It can also compare two identical circuits during this test. Other test instruments can compare identical circuits as well, but can't check the turn-to-turn insulation. EASA estimates that 80 percent of insulation breakdowns start as a turn-to-turn fault.

During the test procedure, this instrument also performs a low-resistance test that (in rare cases) will reveal problems that the surge comparison test

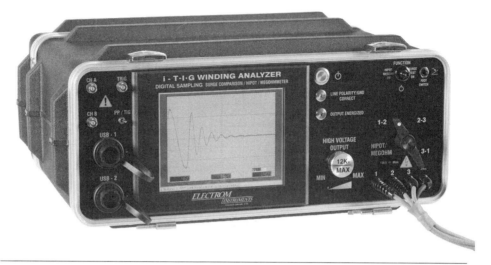

FIGURE 8.9   A surge tester capable of comparison and turn-to-turn voltage stress testing. *Electrom Instruments.*

misses. One example is a poor connection in a multiple-circuit winding. The faulty connection wasn't revealed because the surge test is based on inductance. (When one wire of a multiple wire, winding conductor [for example, when four #12 wires are used instead of one #6] is open, the inductance can be the same, but the resistance is less.)

The surge tester uses up to 10 amperes in its resistance test. The results are printed in hard copy for further review.

The surge tester is often used in predictive maintenance programs. The test unit can be applied at the motor's disconnect, where it will test both the motor and its supply lines. It's not always used during the initial test procedure because of its weight. It is often used to confirm problems that are hard to identify (before a motor is replaced).

The surge tester uses a precisely controlled spike of voltage on the windings. The duration of the spike is so short that it doesn't damage the insulation. The power of the instrument is so well controlled that an arc won't cause carbon tracking, unless the test isn't stopped soon enough.

When a high voltage is impressed turn to turn, any weak spot in the insulation will arc. The test should be stopped as soon as a fault is detected, to prevent further damage. A serious problem has been identified, and the motor can be operated until it's convenient to replace it. There are numerous cases where this tester has saved thousands of dollars in lost production.

Comparison-testing is done with the surge tester's dual-trace (signal) oscilloscope (Fig. 8.10). Two identical windings are tested at the same time. Each signal records the inductance of its respective winding. Any difference in inductance will separate the two signals, indicating a problem. If the signal scrambles, the test is stopped immediately. A scrambled signal indicates arcing caused by a weak spot in the insulation. If the weak spot is visible, it can sometimes be repaired.

Different-shaped signals identify problems such as a turn-to-turn short, phase-to-phase short, wrong connection (internally or externally), and grounded turns.

Rotor Test

With the Electrom rotor bar clamp, (Fig. 8.11) a motor's rotor can be tested in place. Open rotor bars described in Chapter 6 can be detected without disassembling the motor.

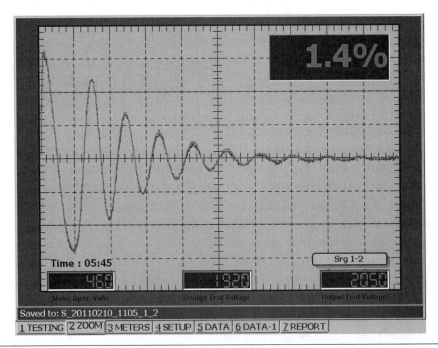

FIGURE 8.10   The signal on the surge tester's dual-trace oscilloscope shows that the two windings being tested have no problem. The motor is being tested from its disconnect. *Electrom Instruments.*

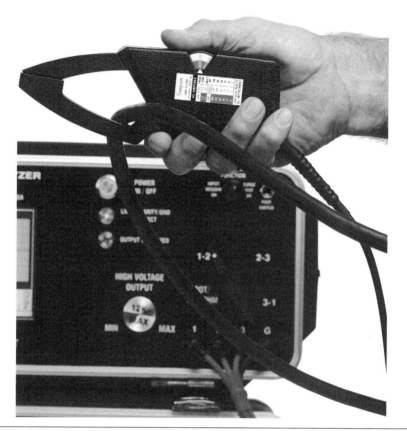

FIGURE 8.11   Rotor clamp being used to test for rotor problems while the motor is running. *Electrom Instruments.*

DC armatures can be tested using a special attachment shown in Fig. 8.12. Large armatures can also be tested, as well as those with equalizer connections.

Polarization Index Test with a Megohmmeter

The polarization index test is done to determine the condition of the insulation that protects the coils. Insulation being tested is located between the coil wire and any path to the frame. (This includes air over the end turns of the coils.) The test is dependable for only form-wound or taped coils that have no wire exposed to air (such as random-wound motors). The random-wound coils are seldom taped and sealed, making the test easily corrupted by changing conditions such as high humidity or conducting dirt.

FIGURE 8.12 Adapter used to test armature windings with a surge tester. *Electrom Instruments.*

In the polarization index test described below, it is assumed that the motor's insulation is not made of epoxy resin or one of the many polymer products. The test doesn't work on these materials.

Many industries don't use this test because it's hard to identify the insulation varieties that invalidate the test. The majority of motors in service are random wound and, as explained above, can't be reliably tested due to the many corrupting variables.

The polarization index test is a trending procedure that documents the degradation of insulation over a timed series of tests. The tests are done at 6- to 12-month intervals. A reading is taken at 1 minute and again at 10 minutes from the time the voltage is applied. The 10-minute reading is divided by the 1-minute reading to get the polarization index. The result should be between 2 and 5. A reading below 2 indicates conducting dirt or moisture. If the result is above 5, the insulation is crystallized (brittle) and could fail.

During the test, the molecules in the insulation are becoming aligned. Current flows and resistance is lower during the alignment of the molecules, and decreases to a lesser value at 10 minutes. The 10-minute value has higher resistance than the 1-minute value.

The motor has to be offline for at least an hour before the polarization index test is performed. Energizing the winding with any type of voltage source will align the free molecules and cause an invalid test. It is good practice to ground the motor leads to the frame for an hour before testing.

As insulation ages, it becomes brittle (crystallized) and there are fewer free molecules. This is why it takes less time to align the free molecules. The test result number becomes higher as the aging progress goes on. At some age, the time it takes to polarize the insulation gives a result greater than 5. At this time the insulation is so brittle that cracks form. If there are conducting contaminants present, the current flows, carbon forms, and there is complete insulation breakdown.

Step Voltage Test

The step voltage test is done with an instrument called a winding analyzer (Fig. 8.13a). The machine is computer-compatible and can give a printout of the test results. The printouts show the voltage and current of each step (Fig. 8.13b).

(a)

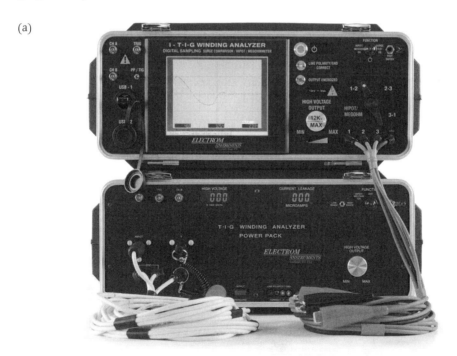

FIGURE 8.13   (a) The winding analyzer used in the step voltage test. *Electrom Instruments.*

(b)

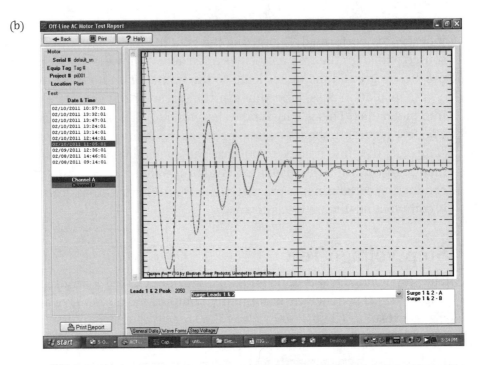

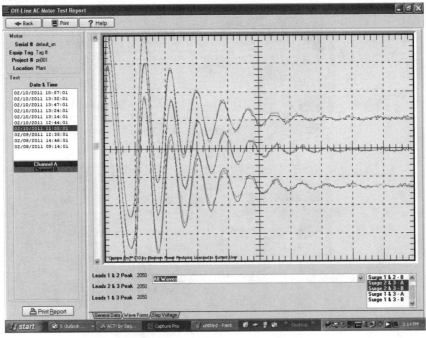

FIGURE 8.13 *(continued)* (b) Printouts of the step voltage test. *Electrom Instruments.*

The test measures the current leakage (caused by contaminants) from the motor's winding to its frame. It is a trending procedure similar to the polarization index test. However, unlike the polarization index test, it works on any type of insulation. It's much more precise than a megohmmeter and can store information for future comparison.

The test can be done at the control terminals or at the motor's junction box. (Control components that can be damaged should be isolated from the test.) Voltage should be raised slowly to its test value to avoid damaging spikes. The leads should be grounded to the frame after the test is completed.

The test procedure is as follows: DC voltage is applied in steps, starting with 500 volts, then 1000 volts, etc., up to the motor's high-potential test value: 2 × rated voltage + 1000 volts.

Each voltage value is applied for 1 minute. The machine then records the voltage and current of each step. There is an inrush of current when the voltage is applied. This current charges the capacitor effect of the winding and polarizes the insulation. (After 1 minute, the current stabilizes to its leakage value.) Leakage current rises as the voltage increases. A sharp increase over the previous test indicates the winding should be cleaned.

High-Potential Test

The high-potential (hi-pot) test is done to stress the phase-to-ground (slot) insulation. Always use an ohmmeter before using the hi-pot test (Fig. 8.14). The windings should read at least 50 megohms to the frame before using the hi-pot. Dirty or wet windings should be cleaned and dried before the hi-pot test. If DC voltage is used, it should be 1.4 times the AC values.

Accepted AC test voltage for high-potential testing is as follows:

- New windings: 2 × rated voltage + 1000 volts for 1 minute
- Small motors rated 250 volts or less: 1000 volts for 1 minute

These variations of the above limits are permissible:

- An equivalent test is 1.2 times the above voltages for 1 second.
- The above voltages × 1.6 for 1 minute (if DC voltage is used).
- Old or contaminated windings: 60 percent of the above voltages for the same time.

FIGURE 8.14 A high-potential (hi-pot) test instrument. *Hipotronics, Inc.*

 AC should never be used on motors over 150 horsepower. Always ground the motor leads to the motor frame after each test to discharge voltage stored by the capacitance of the unit. The hi-pot test can be destructive to the insulation. The test instrument should have an automatic trip feature that shuts it off before damage occurs.

Hermetically sealed refrigeration compressor motors have been damaged when windings were tested while under a vacuum. There are theories as to why this happens, but nothing credible. The best advice is: *Don't hi-pot sealed compressor motors.*

Power Quality Test Instruments

Power quality test instruments are used when electronic devices are suspected of polluting power supply circuits. Some symptoms of line pollution include motors overheating for no apparent reason, premature winding failure, current-caused ball bearing failure, tripped breakers, hot conduit and/or neutral, and damaging voltage spikes. Figure 8.15 shows an instrument for testing power quality.

Circuits which power large numbers of computers are a problem for supply transformers if no correction is done. Electronic motor controls cause

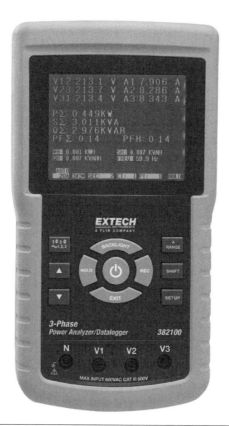

FIGURE 8.15 An instrument for checking a circuit's power factor and power quality. *EXTECH Instruments.*

even greater damage to them than the computers because higher amperes are involved. Remedial equipment (reactors, etc.) is essential for preventing equipment breakdown.

The tests should be done over a long enough period of time to cover all possible conditions.

Identifying and installing remedial equipment is a service that is often outsourced. The use of these test instruments requires specialized training and experience. It is often a one-time service. Safety training for personnel doing these tests is very important because the tests are done with live power.

The meter in Fig. 8.15 is used to determine the phase angle needed for power factor calculating. The consequence of a low power factor can be expensive penalties from the power supplier. A low power factor also affects motors by causing low voltage.

Low power factor is corrected with specially designed capacitors or with synchronous motors. Many industries outsource this procedure because it is normally a one-time service and takes special training. Some electric motor repair centers offer the service. Safety training is very important for personnel doing these tests because they are done with live power lines.

Vibration Sensors

Vibration is the leading cause of bearing failure. It is also an indication of impending bearing failure. Portable instruments work well for spot-checking, but continuously operating machinery should be monitored by attached sensors (Fig. 8.16.)

FIGURE 8.16 This instrument will check vibration, and has noncontact tachometer capability. *EXTECH Instruments.*

Vibration sensors will automatically give a warning when a preset level of vibration is reached. Corrective action can then be scheduled before catastrophic failure occurs. Electric motor repair centers balance repaired motors in accordance with EASA specifications. Some also offer in-plant machinery balancing. The service is often outsourced because it takes special training.

Laser Alignment

Laser alignment is the most accurate of all alignment methods (Fig. 8.17). It has been used successfully on multilevel shaft applications that require exact precision. It is easy to use and virtually eliminates mistakes that are common with other alignment procedures.

FIGURE 8.17   Laser alignment equipment. *SKF USA Inc.*

Flexible couplings are designed for a small degree of misalignment. Experience has shown, however, that vibration will occur unless alignment is nearly perfect.

Particle Analysis

Particle analysis is a service that will predict machinery failure and is usually outsourced. This early-warning procedure has detected problems missed by vibration testing. Both oil and grease analysis can be done with this service.

Oil analysis on a regular basis is recommended. It can identify wear particles of each component in a gearbox. If excess wear of a particular component is detected, the service provider will phone or fax the information immediately.

Contaminants will be identified as well as the condition of the lubricant. This information may indicate the need for more frequent oil changes or a different type of oil (Fig. 8.18).

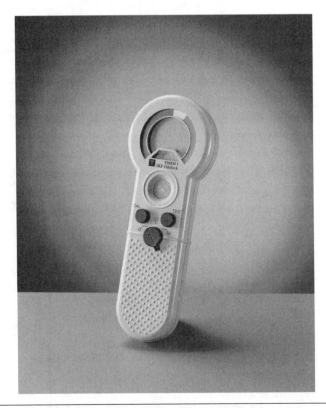

FIGURE 8.18 An oil analysis instrument. *SKF USA Inc.*

Grease analysis will help pinpoint the cause of frequent ball bearing failure. The type of wear (cutting, corrosion, fatigue, etc.) is identified as well as outside contamination. An analysis of the grease's useful life may indicate a need for more frequent lubrication or for a different grade of grease.

Chapter 8 Review

1. Test instruments are vital for accurate troubleshooting (page 367).
   T____ F____

2. Voltage unbalance will cause immediate motor failure (page 369).
   T____ F____

3. The microhmmeter can be used for comparison testing motors larger than 10 horsepower (page 372). T____ F____

4. A tachometer can be used to determine whether a motor is overloaded (page 373). T____ F____

5. The main use for the megohmmeter is to test insulation between the winding and the frame of a motor (pages 374–375). T____ F____

6. An infrared gun (like a voltmeter) should be standard equipment in the toolbox (pages 376–377). T____ F____

7. The electronic stethoscope can be used for preventive maintenance (page 377). T____ F____

8. The surge tester is the only instrument that will reliably stress turn-to-turn insulation (pages 378–379). T____ F____

9. The polarization index test will work on all motors (pages 381–383).
   T____ F____

10. The polarization index test should be done immediately after a motor has been shut off (page 383). T____ F____

11. The step voltage test will work on all motors (page 385). T____ F____

12. The high-potential test voltage should never be applied to a winding when it is set at its full value (pages 385–386). T____ F____

13. The high-potential test is done to check phase-to-phase insulation (page 385). T____ F____

14. 2 × rated voltage + 1000 volts for one minute is acceptable for all motors (page 385). T____ F____

15. What are three signs of power quality problems (pages 386–388)?

16. Give two reasons for improving an industry's power factor (pages 386–388).

17. Vibration analysis and balancing are limited to electric motors (pages 388–389). T___ F___

18. Laser alignment is the most reliable method to reduce coupling vibration (pages 389–390). T___ F___

19. Particle analysis can predict bearing and gear problems in gearboxes (page 390). T___ F___

Index